W0106669

P r e f a c e

This monograph contains, with a few addi-
tions and extensions, my lectures given at the Inter-
national Center for Mechanical Sciences in Udine dur-
ing the month of October, 1969.

Interest in application of the theory of
random processes to mechanics has been growing stead-
ily during the last decade. Indeed, there can be no
doubt that many problems of practical interest call
for a probabilistic interpretation. Frequently, forces,
temperature or other external agencies acting on a
mechanical system cannot be considered to be given
in a deterministic sense. A well known example is the
motion of an airplane through turbulent air. Further-
more, the geometry of a structure may be subject to
random "imperfections", and material properties may
show random deviations from their assumed values.

Obviously, it is impossible to present,
in a relatively short course of a few weeks duration,
all aspects of this highly complex subject. It is my
hope, however, that I have succeeded, to some extent
at least, in alerting my patient audience to a new
field of immense importance, and in paving the way for
them to individual further study.

Chapter I presents those parts of probabi-
lity theory which, in my opinion, constitute the main
tools for application. The chapter can also serve as

---

a handy reference source. Chapter II treats linear
random vibrations and discusses linear differential
equations with random driving functions. Chapter III
is a very brief account of some nonlinear problems,
while in Chapter IV the results of Chapter II and III
are applied to the problem of probabilistic failure
of structures. Chapter V is devoted to stochastic sta
bility and to linear differential equations with para
metric excitation. The final Chapter VI presents a
brief account of optimization of linear systems with
out feedback.

     It is my pleasant duty to record here my
sincere thanks to the authorities of CISM for their
invitation to present these lectures. In particular
and above all, I would like to thank my dear friend
Professor Luigi Sobrero. Without him CISM would not
be in existence.

     My thanks are also due to Prof. W.Olszak,
Rector of CISM, to whom I owe so much. Dr. H.Bargmann,
Dipl. Ing. H.Kastl, Dr. J.L.Zeman and Dr. F.Ziegler
have spent much time and effort on proofreading and
have made many valuable suggestions.

                     H. Parkus

Contents

CHAPTER   I

ELEMENTS OF PROBABILITY THEORY .

The abstract measure - theoretical develop
ment of probability theory during the last three dec-
ades, initiated by Kolmogorov [1] , [2] is, for appli
cations to physical problems, neither necessary nor
even desirable. Therefore, no use is made of it in the
following brief summary of basic definitions and for
mulas of probability theory.

1. Definitions of Probability .

(a) The definition by R. von Mises. Let
an experiment with random outcome be given, and let
it be performed n times.
If, among these n trials, the event E
appears n_E times probability $P[E]$ for E to occur
is defined as the limit of the relative frequency

$$P[E] = \lim_{n \to \infty} \frac{n_E}{n} \qquad (1.1)$$

This definition of probability turns out
to be very well suited, if not essential, for applications since it is based on experiment. It is, there
fore, the one mostly used - explicitly and, sometimes,
implicitly - by engineers and physicists. Nevertheless,
for the axiomatic foundation of probability the defi

nition by Kolmogorov as given below appears to be
superior. One main objection is, of course, that the
limit (1.1) can, in reality, never be determined and
that assuming its existence in fact constitutes already a hypothesis.

(b) The Kolmogorov definition is based on three axioms.
For their formulation some preliminary concepts are
needed.

Let S be the sure (certain) event, appearing at every trial. The set S ("sample space")
contains as elements all possible outcomes of the
trial. A subset $E \subset S$ is called an event. The empty
set $\emptyset$ (which does not contain any outcome at all) is
the impossible event. Denote by $\xi_1, \xi_2 \ldots \xi_n$ the possible
outcomes of the experiment (assumed to be finite for
the time being). Then

$$S = \left\{ \xi_1 , \xi_2 \ldots \xi_n \right\}$$

If at a certain trial the outcome ξ_i appears we say
that those events E occurred which contain ξ_i as element.

The complementary event $\bar{E}$ of E is the
set of elements in S which are not in E. Obviously:
if E occurs $\bar{E}$ can not occur.

The union $E_1 \cup E_2$ occurs if either E_1 or E_2,
or both, occur since $E_1 \cup E_2$ contains all elements that

are in E_1 or in E_2 or in both.

The intersection $E_1 \cap E_2$ occurs if <u>both E_1</u> <u>and E_2</u> occur. If $E_1 \cap E_2 = \emptyset$ the two events are mutual<u>l</u>ly exclusive.

We now define the probability $P[E]$ of an event E as a real number satisfying the following three axioms

(I) $P[E] \geq 0$ "nonnegativity"

(II) $P[S] = 1$ "normalization"

(III) $P\left[\bigcup_{n=1}^{\infty} E_n\right] = \sum_{n=1}^{\infty} P[E_n]$ provided the E_n are mutually exclusive, i.e. $E_i \cap E_j = \emptyset$ for $i \neq j$

"additivity"

From these axioms we conclude immediately

$$P[\bar{E}] = 1 - P[E], \quad P[\emptyset] = 0 \qquad (1.2)$$

Furthermore,

$$P[E_1 \cup E_2] = P[E_1] + P[E_2] - P[E_1 \cap E_2] \qquad (1.3)$$

For a proof we write $E_1 \cup E_2$ and E_2 as the union of two mutually exclusive events

$$E_1 \cup E_2 = E_1 \cup (\bar{E}_1 \cap E_2) \quad , \quad E_2 = (E_1 \cap E_2) \cup (\bar{E}_1 \cap E_2)$$

and use axiom III

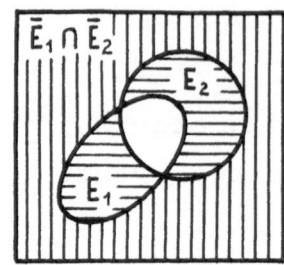 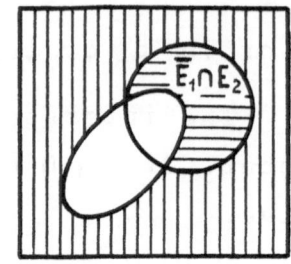

Fig. 1.1.

Putting $E_2 = \bar{E}_1$ in (1.3) one has

$$P\left[E_1 \cup \bar{E}_1\right] = P\left[S\right] = 1$$

It must be emphasized, however, that the converse of $P\left[S\right] = 1$ and $P\left[\emptyset\right] = 0$ is not true! Probability 1 does not mean that the event will certainly occur, and probability 0 does not mean that the event can not occur.

One can easily show that the probability defined by (1.1) satisfies all of the three axioms.

By a fourth axiom we introduce the so-called "conditional probability" $P\left[E \mid E_1\right]$. This is the probability for E occurring if one knows that E_1 will occur

$$(IV) \quad P\left[E \cap E_1\right] = P\left[E_1\right] P\left[E \mid E_1\right], \; P\left[E_1\right] \neq 0 \qquad (1.4)$$

$P\left[E\right]$ is sometimes called "absolute probability".

We call two events E_1 and E_2 (statistically)

---

independent if

$$P\left[E_1 \cap E_2\right] = P\left[E_1\right] P\left[E_2\right] \qquad (1.5)$$

provided $P\left[E_1\right] \neq 0$ and $P\left[E_2\right] \neq 0$. It follows that

$$P\left[E_1 | E_2\right] = P\left[E_1\right] \quad \text{and} \quad P\left[E_2 | E_1\right] = P\left[E_2\right] .$$

If two events are independent so are their complemen tary events.

2. Random Variables .

To each outcome ζ of the experiment we adjoin a real number $x(\zeta)$ such that

(a) the set $\left\{x \leqslant X\right\}$ represents an event E for each real number X , and

(b) the probability of the events $\left\{x = \pm \infty\right\}$ is zero:

$$P\left[x = \pm \infty\right] = 0 \qquad (2.1)$$

$x(\zeta)$ is called random variable.

It follows from property (a) that the set $\left\{X_1 < x \leqslant X_2\right\}$ also represents an event. Indeed

$$\left\{x \leqslant X_1\right\} \cup \left\{X_1 < x \leqslant X_2\right\} = \left\{x \leqslant X_2\right\}$$

The <u>distribution function</u> $F_x(X) \equiv F(X)$
of the random variable x is defined by

$$F(X) = P\left[x \leqslant X\right] \qquad (2.2)$$

Similarly, for a multidimensional random variable
$x = (x_1, x_2 \ldots x_n)$

$$F(X_1, X_2 \ldots X_n) = P\left[x_1 \leqslant X_1, x_2 \leqslant X_2, \ldots x_n \leqslant X_n\right] \qquad (2.3)$$

A random variable is called <u>discrete</u> if
the probability

$$P\left[x = X_i\right] = p(X_i) \equiv p_i \qquad (2.4)$$

is different from zero only on a countable set of val
ues X_i (the <u>spectrum</u> of x), and

$$\sum_{X_i} p(X_i) = 1 \qquad (2.5)$$

The corresponding distribution function is then given
by

$$F(X) = P\left[x \leqslant X\right] = \sum_{X_i \leqslant X} p(X_i) \qquad (2.6)$$

A random variable is called <u>continuous</u>
if its distribution function is continuous and possesses a piecewise continuous derivative $p(X)$

$$\frac{dF}{dX} = p(X),$$

$$P\left[x \leqslant X\right] = F(X) = \int_{-\infty}^{X} p(x)\,dx$$

$$P\left[a < x \leqslant b\right] = \int_{a}^{b} p(x)\,dx$$

$$\int_{-\infty}^{+\infty} p(x)\,dx = F(\infty) = 1$$

$\qquad\qquad (2.7)$

$p(X)$ is called <u>probability density</u>.

The <u>moments</u> of x are defined by

$$m_n\{x\} = \sum_i x_i^n\, p_i \qquad \text{resp.} \qquad m_n\{x\} = \int_{-\infty}^{+\infty} x^n\, p(x)\,dx$$

$\qquad\qquad (2.8)$

The first moment is called expectation (statistical mean), the second moment represents the <u>variance</u>. A short summary follows.

The Most Important Functionals of a Random Variable

Expectation

Discrete

$$m = E\{x\} \equiv <x> = \sum_i x_i\, p_i$$

$$E\{f(x)\} \equiv <f(x)> = \sum_i f(x_i)\, p_i$$

<u>Continuous</u>

$$E\{x\} \equiv \ <x> \ = \int_{-\infty}^{+\infty} x\, p(x)\, dx = m$$

$$E\{f(x)\} \equiv \ <f(x)> \ = \int_{-\infty}^{+\infty} f(x)\, p(x)\, dx$$

E is a linear operator: $E\{x+y\} = E\{x\} + E\{y\}$; $E\{cx\} = c\,E\{x\}$

<u>Variance</u> σ^2

<u>Root mean square</u> σ $\Bigg\}$ $\quad Var\{x\} = \sigma^2 = E\{(x-m)^2\} = E\{x^2\} - m^2$

$$(2.9)$$

<u>Discrete</u>   <u>Continuous</u>

$$\sigma^2 = \sum_i (x_i - m)^2\, p_i \qquad\qquad \sigma^2 = \int_{-\infty}^{+\infty} (x - m)^2\, p(x)\, dx$$

It follows $\qquad E\{(x - X)^2\} = \sigma^2 + (m - X)^2$

<u>Tschebyschev Inequality</u>

$$P\big[|x - m| \ge a\big] \le \frac{\sigma^2}{a^2} \qquad (a > 0) \quad (2.10)$$

<u>Proof:</u>

$$\sigma^2 = \int_{-\infty}^{+\infty} (x-m)^2 p(x)\, dx \ge \int_{|x-m|>a} (x-m)^2 p(x)\, dx \ge a^2 \int_{|x-m|>a} p(x)\, dx =$$

$$= a^2\, P\big[|x - m| \ge a\big]$$

Bienaymé Inequality

$$P\left[|x-a| \geq \varepsilon\right] \leq \frac{E\{|x-a|^n\}}{\varepsilon^n} \tag{2.11}$$

For a proof see $\boxed{4}$, p. 151.

Several Random Variables.   Two Variables u and v.

Joint probability $F(x,y)$

(a) Continuous

$$F(x,y) = P\left[u \leq x, \quad v \leq y\right] = \int\limits_{-\infty}^{x} \int\limits_{-\infty}^{y} p(\xi,\eta) \, d\xi \, d\eta$$

$$p(x,y) = \frac{\partial^2 F(x,y)}{\partial x \partial y}$$

We have

$$P\left[u \leq x\right] = \int\limits_{-\infty}^{x} p(\xi) \, d\xi = \int\limits_{-\infty}^{x} d\xi \int\limits_{-\infty}^{+\infty} p(\xi,\eta) \, d\eta$$

Moments:

$$m_{jk} = E\left\{u^j v^k\right\} = \int\limits_{-\infty}^{+\infty} \int\limits_{-\infty}^{+\infty} x^j y^k \, p(x,y) \, dx \, dy \tag{2.12}$$

In particular

$$m_{2,0} = E\left\{u^2\right\} = \int\limits_{-\infty}^{+\infty} \int\limits_{-\infty}^{+\infty} x^2 \, p(x,y) \, dx \, dy \tag{2.13}$$

$$m_{1,1} = E\{uv\} = \int_{-\infty}^{+\infty}\int_{-\infty}^{+\infty} xy\, p\,(x,y)\, dx\, dy \quad \begin{array}{l}\text{Crosscorrelation} \\ \text{function}\end{array} \quad (2.14)$$

$$m_{0,2} = E\{v^2\} = \int_{-\infty}^{+\infty}\int_{-\infty}^{+\infty} y^2 p\,(x,y)\, dx\, dy \qquad\qquad (2.15)$$

$$E\{f(u,v)\} = \int_{-\infty}^{+\infty} dx \int_{-\infty}^{+\infty} f\,(x,y)\, p\,(x,y)\, dy \qquad\qquad (2.16)$$

(b)   Discrete

$$P\left[u = x_i,\ v = y_j\right] = p_{ij} \quad , \quad \sum_i \sum_j p_{ij} = 1 \qquad (2.17)$$

$$P\left[u = x_i\right] = \sum_j P\left[u = x_i,\ v = y_j\right] = \sum_j p_{ij} = p_i \qquad (2.18)$$

Also $\quad P\left[v = y_j\right] = \sum_i p_{ij} = q_j \qquad\qquad\qquad\qquad (2.19)$

Conditional Probability

$$P\left[v = y_j|\ u = x_i\right] = \frac{P\left[u = x_i, v = y_j\right]}{P\left[u = x_i\right]} = \frac{p_{ij}}{p_i} \qquad (2.20)$$

Moments

$$m_{k\ell} = E\left\{u^k v^\ell\right\} = \sum_i \sum_j x_i^k y_j^\ell p_{ij} \qquad (2.21)$$

In particular

$$m_{2,0} = E\left\{u^2\right\} = \sum_i \sum_j x_i^2 p_{ij} \quad , \quad m_{0,2} = E\left\{v^2\right\} = \sum_i \sum_j y_j^2 p_{ij}$$

$$(2.22)$$

$$m_{1,1} = E\left\{uv\right\} = \sum_i \sum_j x_i y_j p_{ij}$$

Similarly for n random variables with

$$P\left[u_1 = x_1^{(i)}, u_2 = x_2^{(j)} \ldots u_n = x_n^{(s)}\right] = p_{i,j\ldots s}$$

Characteristic Function $\varphi(q) = E\left\{e^{iqx}\right\}$

$$\varphi(q) = \sum_j e^{iqx_j} \, p(x_j) \qquad \varphi(q) = \int_{-\infty}^{+\infty} e^{iqx} p(x) dx \quad (2.23)$$

where q is a real number, $-\infty < q < +\infty$.

The probability distribution is uniquely
defined by the characteristic function:

$$p(x_i) = \lim_{Q \to \infty} \frac{1}{2Q} \int_{-Q}^{+Q} e^{-iqx_i} \varphi(q) dq , \qquad p(x) = \frac{1}{2\pi} \int_{-\infty}^{+\infty} e^{-iqx} \varphi(q) dq \quad (2.24)$$

Therefore, in the continuous case, $p(x)$ is the Fourier
transform of $\varphi(q)$.

For a multidimensional random variable,

Eq (2.3), we have

$$\varphi(q_1, q_2 \ldots q_n) = E\left\{\exp\, i\, (q_1\, x_1 + q_2\, x_2 + \ldots + q_n\, x_n)\right\} =$$

$$= \sum_r e^{i\, q_1\, x_1^{(r)}} \sum_r e^{i\, q_2 x_2^{(r)}} \ldots \sum_r e^{i\, q_n x_n^{(r)}}\, p\, (x_1^{(r)}, x_2^{(r)} \ldots x_n^{(r)}) \qquad (2.25)$$

$$= \int_{-\infty}^{+\infty} \ldots \int_{-\infty}^{+\infty} \exp\left[i\, (q_1\, x_1 + \ldots q_n x_n)\right] p\, (x_1, \ldots x_n)\, dx_1 \ldots dx_n$$

where $\left\{x_i^{(1)}, x_i^{(2)} \ldots\right\}$ represents the spectrum of the discrete random variable x_i .

Jointly distributed (multidimensional) random variables $x_1, x_2 \ldots x_n$ are called <u>independent</u>, cf. Eq. (1.5), if

$$P\left[x_1 \leq X_1, x_2 \leq X_2 \ldots x_n \leq X_n\right] = P\left[x_1 \leq X_1\right] P\left[x_2 \leq X_2\right] \ldots P\left[x_n \leq X_n\right] =$$

$$= F(X_1)\, F(X_2) \ldots F(X_n) \qquad (2.26)$$

and, hence, also

$$p(X_1, \ldots X_n) = p(X_1)\, p(X_2) \ldots p(X_n) \qquad (2.27)$$

For the characteristic function one has

$$\varphi(q_1, q_2 \ldots q_n) = \varphi(q_1)\, \varphi(q_2) \ldots \varphi(q_n) \qquad (2.28)$$

and for the expectations, with arbitrary functions
$f_1(x_1) \ldots f_n(x_n)$,

$$E\left\{f_1(x_1)\ldots f_n(x_n)\right\} = E\left\{f_1(x_1)\right\} \ldots E\left\{f_n(x_n)\right\} \qquad (2.29)$$

Two jointly distributed random variables x_1 and x_2 are called <u>uncorrelated</u> if their <u>covariance</u> $Cov\left\{x_1, x_2\right\}$ vanishes, where

$$Cov\left\{x_1, x_2\right\} \equiv \lambda_{12} = E\left\{(x_1 - m_1)(x_2 - m_2)\right\} =$$
$$\qquad\qquad\qquad (2.30)$$
$$= E\left\{x_1 \cdot x_2\right\} - m_1 m_2$$

Frequently, the correlation coefficient ϱ_{12} is used

$$\varrho_{12} = \frac{\lambda_{12}}{\sigma_1 \sigma_2} = E\left\{\frac{x_1 - m_1}{\sigma_1} \frac{x_2 - m_2}{\sigma_2}\right\} \qquad (2.31)$$

which is then zero for uncorrelated variables.

Two independent variables are also uncorrelated. The converse is not necessarily true.

If, for two variables x and y ,

$$E\left\{xy\right\} = 0 \qquad (2.32)$$

they are called orthogonal.

The probability distribution of the sum $z = x + y$ of two <u>independent</u> random variables x and y, with densities $f(X)$ and $g(Y)$, is given by the convolution integral

$$p(Z) = \int_{-\infty}^{+\infty} f(x) g(Z-x) \, dx = \int_{-\infty}^{+\infty} f(Z-x) g(x) \, dx \qquad (2.33)$$

<u>Proof</u>: The characteristic function of z is, with (2.29),

$$\varphi_z(q) = E \left\{ e^{iq(x+y)} \right\} = \varphi_x(q) \, \varphi_y(q)$$

Hence, from (2.24) and (2.23)

$$p(Z) = \frac{1}{2\pi} \int_{-\infty}^{+\infty} e^{-iqZ} \varphi_x(q) \, \varphi_y(q) \, dq = \frac{1}{2\pi} \int_{-\infty}^{+\infty} e^{-iqZ} \varphi_y(q) \int_{-\infty}^{+\infty} e^{iqx} f(x) \, dx \, dq =$$

$$= \frac{1}{2\pi} \int_{-\infty}^{+\infty} f(x) \int_{-\infty}^{+\infty} e^{-iq(Z-x)} \varphi_y(q) \, dq \, dx = \int_{-\infty}^{+\infty} f(x) \, g(Z-x) \, dx$$

Additional Remarks

(a) If x and y are <u>normally</u> distributed, $N(m, \sigma^2)$ and $N(n, \tau^2)$, then z is normal with mean $m + n$ and variance $\sigma^2 + \tau^2$.

(b) If x and y have <u>binomial</u> distributions with parameters n , p and m , p ,respectively, then z is binomial with parameters $n+m$, p .

(c) If x and y are <u>Poisson</u> distributed with mean values λ_x and λ_y, respectively, then z is Poisson distributed with mean $\lambda_x + \lambda_y$.

(d) If x and y are <u>negative binomial</u> distributed with parameters k_x , p and k_y , p , respec tively, then z is negative binomial with parameters $k_x + k_y$, p .

(e) If x and y are <u>gamma</u> distributed with parameters n , λ and m , λ , respectively, then z is gamma distributed with parameters $n+m,\lambda$.

<u>Central Limit Theorem</u>. Let x_1 , $x_2 \ldots x_n$ be <u>independ-ent</u> random variables with arbitrary probability distributions, mean values m_i and variances σ_i^2 . The sum

$$y = \sum_{i=1}^{n} x_i$$

is then normal distributed with mean

$$m = \sum_{i=1}^{n} m_i$$

and variance

$$\sigma^2 = \sum_{i=1}^{n} \sigma_i^2$$

for $n \longrightarrow \infty$ provided two positive numbers a and b exist such that

$$E\left\{|x_i|^{2+a}\right\} < b\sigma_i^2 \qquad \text{for all } i = 1,2 \ldots$$

Multidimensional   Distributions

Two-dimensional Normal Distribution

$$p(x,y) = \frac{1}{2\pi\sigma\tau\sqrt{1-\varrho^2}} \exp\left\{\frac{-1}{2(1-\varrho^2)}\left[\left(\frac{x-m}{\sigma}\right)^2 - 2\varrho\frac{x-m}{\sigma}\frac{y-n}{\tau} + \left(\frac{y-n}{\tau}\right)^2\right]\right\}$$

$$(2.34)$$

where

$$m = E\{x\}, \ n = E\{y\}, \ \sigma^2 = E\{x^2\} - m^2, \ \tau^2 = E\{y^2\} - n^2, \ \varrho\sigma\tau = E\{xy\} - mn$$

For x , y independent one has $p(x,y) = p(x)\,p(y)$
and, hence, $\varrho = 0$.

N-dimensional Normal Distribution.

$$p(x_1...x_n) = \frac{1}{\sqrt{(2\pi)^n|\lambda_{ik}|}} \exp\left[-\frac{1}{2}\sum_{i=1}^{n}\sum_{k=1}^{n}\mu_{ik}(x_i-m_i)(x_k-m_k)\right]$$

$$(2.35)$$

where

$$\lambda_{ik} = \lambda_{ki} = E\left\{(x_i-m_i)(x_k-m_k)\right\} = \begin{cases} \sigma_i^2 & \text{for } i=k \\ \\ E\{x_i x_k\} - m_i m_k = \varrho_{ik}\sigma_i\sigma_k & \text{for } i\neq k \end{cases}$$

form the $n \times n$ matrix of the central moments, $|\lambda_{ik}|$

is the corresponding determinant, and μ_{ik} are the elements of the inverse matrix.

N-dimensional Poisson Distribution.

$$p\left(x_1, x_2 \ldots x_n\right) = e^{-\left(\lambda_1 + \lambda_2 + \ldots + \lambda_n\right)} \frac{\lambda_1^{x_1} \lambda_2^{x_2} \ldots \lambda_n^{x_n}}{x_1! \, x_2! \ldots x_n!} \qquad (2.36)$$

Conditional Expectation, Conditional Variance and Conditional Distribution of a Normal Random Variable.

Let x and y be two <u>normal</u> random variables with zero expectations and let $z = y - \alpha x$ be orthogonal to x :

$$E\left\{(y - \alpha x)\, x\right\} = 0$$

whence

$$\alpha = \frac{E\{xy\}}{E\{x^2\}} \qquad (2.37)$$

Since z too is normal with zero mean, x and z are not only orthogonal but also uncorrelated and, in addition, independent, cf. (2.34). Hence

$$E\left\{y - \alpha x \mid x\right\} = E\left\{y - \alpha x\right\} = 0$$

But [*)]

$$E\{y - \alpha x | x\} = E\{y | x\} - \alpha E\{x | x\} = E\{y | x\} - \alpha x$$

Therefore

$$E\{y | x\} = \alpha x \qquad\qquad (2.38)$$

with α from (2.37).

For the determination of the condition-al variance one has first

$$Var\{y | x\} = Var\{z | x\} = E\{(y - \alpha x)^2 | x\} = E\{(y - \alpha x)^2\}$$

since $y - \alpha x$ is independent of x . Expansion, taking (2.37) into account, renders

$$Var\{y | x\} = \sigma_{y|x}^2 = E\{y^2\} - \frac{E^2\{xy\}}{E\{x^2\}} \qquad (2.39)$$

---

*) The conditional probability of a random variable is, by extending axiom (1.4), defined as

$$F(X | E) = P[x \le X | E] = \frac{P[(x \le X) \cap E]}{P[E]}$$

provided

$$P[E] \neq 0$$

or, using the notation of (2.34)

$$\sigma^2_{y|x} = \left(1 - \varrho^2\right) \tau^2 \;, \quad \alpha = \varrho \, \frac{\tau}{\sigma} \qquad (2.40)$$

The conditional probability density is, therefore,

$$p(y|x) = \frac{1}{\tau \sqrt{2\pi\left(1-\varrho^2\right)}} \; \exp\left[-\frac{\left(y-\alpha x\right)^2}{2\left(1-\varrho^2\right)\tau^2} \right] \qquad (2.41)$$

Functions of One Random Variable.

       In order to determine the probability distribution $P\left[y \leqslant Y\right]$ of a function $y = f(x)$ of the random variable x from $P\left[x \leqslant X\right]$, equation $y = f(x)$ has to be solved for x.

Example $y = \dfrac{1}{x^2} > 0$

From

$$x_1 = -\frac{1}{\sqrt{y}} \;, \quad x_2 = +\frac{1}{\sqrt{y}}$$

one has at once

$$P\left[y \leqslant Y\right] = P\left[x \leqslant -\frac{1}{\sqrt{Y}}\right] + P\left[x \geqslant \frac{1}{\sqrt{Y}}\right] = F_x\left(-\frac{1}{\sqrt{Y}}\right) + 1 - F_x\left(\frac{1}{\sqrt{Y}}\right)$$

for $Y \geqslant 0$ and $P\left[y \leqslant Y\right] = 0$ for $Y < 0$

       A general formula can be given for the proba

bility density $q(y)$. Let $x_1, x_2 \ldots x_n \ldots$ be all <u>real</u>
roots of the equation $f(x) = y$. Then

$$q(y) = \sum_{i=1}^{\infty} \frac{p(x_i)}{|f'(x_i)|} \qquad (2.42)$$

<u>Proof:</u>

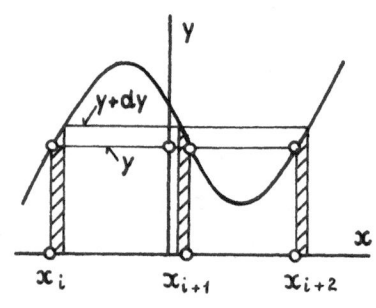

Fig. 1.2

$$q(y)\,dy = P\left[y < Y < y + dy\right] =$$

$$= \ldots P\left[x_i < X < x_i + |dx_i|\right] +$$

$$+ P\left[x_{i+1} < X < x_{i+1} + |dx_{i+1}|\right] + \ldots =$$

$$= p(x_i)|dx_i| + p(x_{i+1})|dx_{i+1}| + \ldots$$

With $f'(x_i)\,dx_i = dy$ Eq. (2.42) follows.

<u>Functions of Several Random Variables.</u>

Let
$$y_1 = f_1(x_1, \ldots x_n), \ldots y_k = f_k(x_1, \ldots x_n) \qquad (2.43)$$

We look for the distribution $q(y_1, \ldots y_k)$.

If $k < n$ the number k of the y_i is first
supplemented to n by introducing the auxiliary variables

$$y_{k+1} = x_{k+1}, \ldots y_n = x_n$$

Then we solve the system

$$f_1\left(x_1, \ldots x_n\right) = y_1$$

$$\cdot \quad \cdot \quad \cdot \quad \cdot \quad \cdot \quad \quad \cdot$$

$$f_n\left(x_1, \ldots x_n\right) = y_n$$

for the x_i . If it has <u>one</u> real solution $x_1 \ldots x_n$, then

$$q\left(y_1, \ldots y_n\right) = \frac{p\left(x_1 \ldots x_n\right)}{\left|J\left(x_1 \ldots x_n\right)\right|} \qquad (2.44)$$

where J represents the Jacobian of the system (2.43)

$$J\left(x_1 \ldots x_n\right) = \begin{vmatrix} \dfrac{\partial f_1}{\partial x_1} & \cdots & \dfrac{\partial f_1}{\partial x_n} \\ \cdot & \cdots & \cdot \\ \dfrac{\partial f_n}{\partial x_1} & \cdots & \dfrac{\partial f_n}{\partial x_n} \end{vmatrix} \qquad (2.45)$$

If the system has more than one real solution the right hand side of (2.44) has to be replaced by

$$\sum \frac{p\left(x_{i1} \ldots x_{in}\right)}{\left|J\left(x_{i1}, \ldots x_{in}\right)\right|}$$

If the system has no real solution then

$$q\left(y_1, \ldots y_n\right) = 0$$

3. Random Processes .

Again, let an experiment be given characterized by the results (outcomes, elementary events)
ζ forming the space $S\left\{\zeta_1,\zeta_2...\zeta_n\right\}$. Probabilities
$P\left[E\right]$ are assigned to certain subsets E of S . However, to each outcome of a trial we now adjoin not a
real number x but a (real or complex) <u>time function</u>

$$x\left(t,\zeta\right)$$

This family of functions forms a <u>stochastic process</u>.
In the following we shall use the notation $x(t)$ omitt-
ing the dependence on ζ. Of course, x can be a multidimensional variable.

From the above we see that $x\left(t\right)$ may
have four different meanings:

   (1) A family of time functions (t and ζ variable)

   (2) A single time function (t variable, $\zeta = \zeta_i$ fixed)

   (3) A random variable ($t=t_n$ fixed, ζ variable)

   (4) A single number (t and ζ fixed)

Random functions of a stochastic pro-
cess may mean quite different things. A realization
of the one-dimensional Brownian motion, for instance, will perhaps look as in Fig. 1.3. It is continuous but nowhere differentiable.

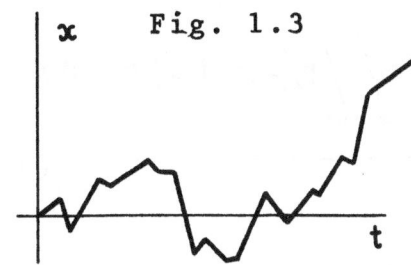

x   Fig. 1.3

Above all, however: if $x\left(t,\xi_i\right)$ is known for $t < T$ its further course can not be predicted.

On the other hand, if one considers the random process

cess

$$x\left(t,\zeta\right) = a(\zeta)\cos\left[\omega(\zeta)t+\varepsilon(\zeta)\right]$$

then each realization $x\left(t,\zeta=\xi_i\right)$ represents an entirely regular curve, which is known for $t > T$ once it is known for $t \leqslant T$.

Probability distribution $F\left(X;t\right) = P\left[x(t)\leqslant X\right]$ and density $p\left(X;t\right) = \partial F/\partial X$ are now, in general, functions of time t . However, they are no longer sufficient to describe the process. In the sense of probability theory a random process is to be considered as being determined if its n-dimensional (n=1,2,3..) probability distributions or probability densities are given (each higher distribution contains all preceding ones and describes the process with increasing accuracy):

One-dim.: $P_1\left[x(t)\leqslant \xi\right] = \int_{-\infty}^{\xi} p_1\left(x;t\right)dx$, $P_1 = 1$ for $\xi = \infty$

Two-dim.: $P_2 \left[x(t) \leqslant \xi , x(s) \leqslant \eta \right] = \int_{-\infty}^{\xi} \int_{-\infty}^{\eta} P_2(x,y;t,s) \, dx \, dy$

- (3.1)

n-dim.:

$P_n \left[x(t_1) \leqslant \xi_1 , x(t_2) \leqslant \xi_2 \ldots x(t_n) \leqslant \xi_n \right] = \text{"joint probability"}$

$= \int_{-\infty}^{\xi_1} \int_{-\infty}^{\xi_2} \ldots \int_{-\infty}^{\xi_n} P_n(x_1, \ldots x_n ; t_1, \ldots t_n) \, dx_1 \ldots dx_n$

We have

$P_{n-1} \left(\xi_1 , \xi_2 , \ldots \xi_{n-1} ; t_1 , t_2 \ldots t_{n-1} \right) = \lim_{\xi_n \to \infty} P_n$

Connected with the probability law for the n random variables $x(t_1) \ldots \ldots x(t_n)$ is the joint characteristic function

$\varphi (q_1 , q_2 \ldots q_n ; t_1 , t_2 \ldots t_n) = E \left\{ \exp \left[i \left(q_1 x(t_1) + \ldots + q_n x(t_n) \right) \right] \right\} =$

$= \int_{-\infty}^{+\infty} \ldots \int_{-\infty}^{+\infty} \exp \left[i \left(q_1 x_1 + \ldots q_n x_n \right) \right] P_n(x_1, \ldots x_n; t_1, \ldots t_n) dx_1 \ldots dx_n$

(3.2)

Mean value $E \left\{ x(t) \right\} = m(t) = < x(t) > = \int_{-\infty}^{+\infty} x p(x;t) dx$ (3.3)

Correlation Function (Autocorrelation)

$$E\left\{x(t)\cdot x(s)\right\} = R(t,s) = <x(t)\cdot x(s)> = \int\limits_{-\infty}^{+\infty}\int\limits_{-\infty}^{+\infty} xy\, p(x,y;t,s)\, dx\, dy \qquad (3.4)$$

and similarly for the higher moments. For the variance we have

$$\sigma^2(t) = R(t,t) - m^2(t) \qquad (3.5)$$

Discrete Distribution

$$P_1\left[x(t) = X_i\right] = p_i(t)$$

$$\qquad (3.6)$$

$$P_2\left[x(t) = X_i,\, x(s) = Y_j\right] = p_{ij}(t,s)$$

- - - - - - - - - - - -

$$P_1\left[x(t) \leqslant \xi\right] = \sum_{X_i \leqslant \xi} p_i(t)$$

$$\qquad (3.7)$$

$$P_2\left[x(t) \leqslant \xi,\, x(s) \leqslant \eta\right] = \sum_{X_i \leqslant \xi}\sum_{Y_j \leqslant \eta} p_{ij}(t,s)$$

- - - - - - - - - - - - - - -

$$P_n\left[x(t_1) \leqslant \xi_1 \ldots x(t_n) \leqslant \xi_n\right] = \sum_{X_i \leqslant \xi_1} \ldots \sum_{Z_j \leqslant \xi_n} p_{ij\ldots n}\left(t_1, t_2 \ldots t_n\right)$$

Expectation $\quad E\left\{x(t)\right\} = \sum_i x_i\, p_i(t)$ $\qquad$ (3.8)

Autocorrelation

$$R(t,s) = E\left\{x(t)x(s)\right\} = \sum_i \sum_j x_i\, y_j\, p_{ij}(t,s) \qquad (3.9)$$

In the case of a <u>normal distribution</u>
the random process is completely determined if $m(t)$
and $R(t,s)$ are known since then all $p\left(\xi_1...\xi_n\,;t_1...t_n\right)$
are determined, cf. (2.35).

<u>Stochastic Continuity, Differentiation and Integration.</u>

In order to apply the rules of analysis
to random functions [*] the concepts of continuity and
limit have to be redefined. This can be done in various ways. Some definitions are given below.

---

[*] The difficulty arises because a stochastic process
represents a <u>family</u> of functions. Some members of the
family may behave quite reasonably while others may
be pathological. Our definition shall, however, be
valid for the <u>entire</u> family!

(a) Ordinary (deterministic) continuity

$$\lim_{h \to 0} \left[x\,(t+h) - x(t) \right] = 0$$

for each realization ("sample") of the process.

(b) Almost sure continuity

$$\lim_{h \to 0} \left[x\,(t+h) - x(t) \right] = 0$$

with probability $P = 1$.

(c) Continuity in mean square

$$\lim_{h \to 0} E \left\{ \left| x\,(t+h) - x(t) \right|^2 \right\} = 0$$

Definitions (a) and (b) refer to sample properties and, hence, are more desirable in applications than (c). However, definition (a) is too restrictive and is not being used in probability theory. (c) is the most commonly used form. (b) and (c) do not imply each other.

In the same manner derivative and Riemann integral of a random function may be defined. For instance, for the mean square derivative we have

$$\lim_{h \to 0} E \left\{ \left[\frac{x(t+h) - x(t)}{h} - \dot{x}(t) \right]^2 \right\} = 0$$

Theorem. For the existence of $\dot{x}(t)$ in the above sense the existence of

$$\frac{\partial R}{\partial t} \, , \, \frac{\partial R}{\partial s} \, , \, \frac{\partial^2 R}{\partial t \, \partial s}$$

at $t = s$ is necessary and sufficient. $R(t,s)$ is the correlation function of the process. We give a proof for the stationary case. It suffices to show the existence of

$$E\left\{\left[\frac{x(t+h)-x(t)}{h}\right]^2\right\} = \frac{1}{h^2}\left[R(0)-2R(h)+R(0)\right] = 2\,\frac{R(0)-R(h)}{h^2}$$

for $h \to 0$. For this, in turn, the existence of

$$\lim_{h \to 0} \frac{R(h)-R(0)}{h} = R'(0) = 0$$

and

$$\lim_{h \to 0} 2\,\frac{R(h)-R(0)}{h^2} = R''(0)$$

is necessary and sufficient. But, as a consequence of the symmetry of $R(t)$ the first condition is satisfied if $R'(0)$ exists. This completes the proof which, for a nonstationary process may be given in an analogous manner.

For a differentiable process

$$E\left\{\dot{x}(t)\right\} = \frac{d}{dt} E\left\{x(t)\right\} \tag{3.10}$$

"Expectation and differentiation commute".

Similarly, if the Riemann double integral
$\int_a^b \int_a^b R(t,s)\,dt\,ds$ exists in the square (a,b), then $\int_a^b x(t)\,dt$
exists in mean square, and

$$E\left\{\int_a^b x(t)\,dt\right\} = \int_a^b E\left\{x(t)\right\}\,dt \tag{3.11}$$

"Expectation and differentiation commute".

Stationary Random Processes.

If m is independent of t, and if $R(t,s)$ depends on the difference $t-s$ only the random process is called (weakly) stationary. σ^2 is then a constant. We have

$$R(t-s) = E\left\{x(t)x(s)\right\} = R(s-t) = R(\tau),\ \tau = |s-t| \tag{3.12}$$

and

$$\sigma^2 = R(0) - m^2 \tag{3.13}$$

Note. If t denotes a space variable rather than time the process is called homogeneous.

Note. The process is called stationary in the strict sense if its entire statistics remains unchanged under a time-shift

$$p_n\left(x_1,\ldots x_n;\, t_1,\ldots t_n\right) = p_n\left(x_1,\ldots x_n;\, t_1+\tau,\ldots t_n+\tau\right)$$

$$(3.14)$$

for each n . It follows that

$$p_1\left(x;t\right) = p_1\left(x\right), \qquad E\left\{x\left(t\right)\right\} = m = \text{const}$$

$$(3.15)$$

$$p_2\left(x_1,x_2;\, t_1,t_2\right) = p_2\left(x_1,x_2;\, \tau\right),$$

$$E\left\{x\left(t+\tau\right).x\left(t\right)\right\} = R\left(\tau\right) = R\left(-\tau\right)$$

The same holds true for the cross-correlation of a two-dimensional process $x\left(t\right), y\left(t\right)$:

$$R_{xy}\left(\tau\right) = E\left\{x\left(t+\tau\right).y\left(t\right)\right\} \qquad (3.16)$$

If (3.14) holds for $n \leqslant k$ only rather than for all n the process is called <u>stationary of order k</u> . A process of order 2 is also weakly stationary; the converse is not true!

4. Transformation of a Random Process .

Let an operator T be defined on the set of random functions $x(t)$, and let T adjoin to each function of this set one and only one function $y(t)$:

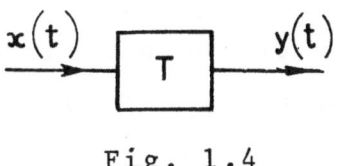

Fig. 1.4

$$y(t) = T\left[x(t)\right] \quad (4.1)$$

We speak of a transforma-
tion or mapping. Frequently, $x(t)$ and $y(t)$ are called
"input" and "output", respectively, of the system defined by T.

Since, to each outcome ζ_i there corresponds one time-
-function $x(t,\zeta_i)$ and, hence, one time-function $y(t,\zeta_i)$
the operator is deterministic. If, however, two outcomes ζ_1 and ζ_2 exist such that

$$x\left(t,\zeta_1\right) = x\left(t,\zeta_2\right) \quad \text{but} \quad y\left(t,\zeta_1\right) \neq y\left(t,\zeta_2\right)$$

we speak of a random operator. If, for instance, T is
a differential operator then it is deterministic or
random depending on whether its coefficients are deterministic or random, respectively. For the time being we restrict ourselves to deterministic operators[*]
and consider two special cases.

(a) Time - Invariant Systems. Let the operation $T(x)$
be a function of x but not of $\dot{x}$ or $\int_0^t x(\tau)\,d\tau$ etc.

Then the output $y(t_1)$ at time t_1 depends on $x(t_1)$

---

[*] Random operators will be discussed in Chapter $\underline{V}$.

only and is independent of the past or future values of $x(t)$: the system has no memory.

Since now $y = f(x)$ the p.-density of the process $y(t)$ may be determined from (2.42). Similarly, one finds the joint density $\quad q_n(y_1, \ldots y_n; t_1, \ldots t_n) \quad$ from (2.44), with $\quad y_i = y(t_i) = f\left[x(t_i)\right]$.

For expectation and autocorrelation we have

$$E\left\{y(t)\right\} = \int_{-\infty}^{+\infty} f(x)\, p(x;t)\, dx \qquad (4.2)$$

$$R_{yy}(t_1, t_2) = E\left\{y(t_1) \cdot y(t_2)\right\} = \int_{-\infty}^{+\infty}\int_{-\infty}^{+\infty} f(x_1)\, f(x_2)\, p_2(x_1, x_2; t_1, t_2)\, dx_1\, dx_2$$

$$(4.3)$$

If the input $x(t)$ is stationary in the strict sense or stationary of order n then the output $y(t)$ is stationary in the strict sense or stationary of order n , respectively. The proof follows from a consideration of $q_n(y_1, \ldots y_n; t_1, \ldots t_n)$ since the Jacobian is independent of t.

If $x(t)$ is only weakly stationary then $y(t)$ is not necessarily stationary in any sense.

(b) <u>Linear System</u>. If the operator $T \equiv L$ has the property that

$$L\left[\Sigma\, a_i\, x_i(t)\right] = \Sigma\, a_i\, L\left[x_i(t)\right] \qquad (4.4)$$

for arbitrary a_i and $x_i(t)$ then it is called linear.

We find, as a consequence of (3.10) and (3.11), that

$$E\left\{L\left[x(t)\right]\right\} = L\left[E\left\{x(t)\right\}\right] \qquad (4.5)$$

"Expectation and linear operator commute".

For the autocorrelation we have

$$\left.\begin{aligned} R_{xy}(t,s) &= L_s\left[R_{xx}(t,s)\right] \\[2mm] R_{yy}(t,s) &= L_t\left[R_{xy}(t,s)\right] = L_t\, L_s\left[R_{xx}(t,s)\right] \end{aligned}\right\} \qquad (4.6)$$

where L_t and L_s indicate that s and t, respectively, are parameters only.

An important linear operator is the polynomial $P(D)$, with constant coefficients, in the derivative $D \equiv d/dt$:

$$P(D) = a_n D^n + a_{n-1} D^{n-1} + \ldots + a_o \qquad (4.7)$$

A second, frequently occurring linear transformation is of the form

$$y(t) = \int_{-\infty}^{t} h(t-\tau)\, x(\tau)\, d\tau \qquad (4.8)$$

where $h(t,\tau)$ denotes a deterministic function, the so-

-called "impulse response" of the system.

It is important to note that the application of a linear (deterministic) operator to a normal random function again produces a normal function. The reason is in the fact that the sum of (dependent or independent) normal random functions is again normal.

5. Correlation Theory .

A random process may be considered to be completely determined if all its p.-distributions (3.1) are known. However, the determination of these distributions is, in general, a difficult or even impossible task. Frequently one has, therefore, to be satisfied with the knowledge of mean value and correlation function.   This highly restricted treatment of random processes is called correlation theory.

Some Important Properties of Correlation Functions.

For the sake of generality we admit random functions that are complex - valued

$$x(t) = u(t) + iv(t)$$

Then, with [*] denoting the complex-conjugate,

$$R(t,s) = E\left\{x(t) \cdot x^*(s)\right\} \qquad (5.1)$$

and we have

$$(I) \qquad R(t,s) = R^*(s,t) \qquad (5.2)$$

For a real random function this becomes symmetry:

$$R(t,s) = R(s,t)$$

Consider now the expression $|x(t) \pm x(s)|^2$, which is real and nonnegative. Hence, for its expecta̲tion we have

$$E\left\{|x(t) \pm x(s)|^2\right\} \geqslant 0$$

or, after expanding:

(II) $R(t,t) + R(s,s) \geqslant 2|\mathcal{R}e\, R(t,s)|$ (5.3)

For a stationary function $R(t,s) \rightarrow R(t-s) = R(\tau)$, and (I) and (II) go over into

(Ia) $R(\tau) = R^*(-\tau)$

            $R(\tau) = R(-\tau)$ (real)   (5.2a)

(IIa) $R(0) \geqslant |\mathcal{R}e\, R(\tau)|$

            $R(0) \geqslant R(\tau)$ (real)   (5.3a)

For the cross-correlation of two complex random functions $x(t)$ and $y(t)$

$$R_{xy}(t,s) = E\left\{x(t)\, y^*(s)\right\}$$ (5.4)

analogous relations hold

(III) $R_{xy}(t,s) = R_{yx}^*(s,t)$

            $R_{xy}(t,s) = R_{yx}(s,t)$ (real)   (5.5)

$$\left(\overline{\text{IV}}\right) \qquad |R_{xy}(t,s)| \leqslant \sqrt{R_x(t,t)\ R_y(s,s)} \qquad\qquad (5.6)$$

The last relation follows from

$$|R_{xy}(t,s)| = \int\limits_{-\infty}^{+\infty}\int\limits_{-\infty}^{+\infty} xy\, p\,(x,y\,;t,s)\ dx\ dy$$

with the aid of the Schwarz inequality $\boxed{8}$, p. 46.

In the stationary case, $R_{xy}(t,s) \rightarrow R_{xy}(t-s) = R_{xy}(\tau)$, equations (5.5) and (5.6) simplify to

$$\left(\overline{\text{III}}\,a\right) \qquad R_{xy}(\tau) = R_{yx}^{*}(-\tau)$$

$$(5.5a)$$

$$R_{xy}(\tau) = R_{yx}(-\tau) \quad (\text{real})$$

$$\left(\overline{\text{IV}}\,a\right) \qquad |R_{xy}(\tau)| \leqslant \sqrt{R_x(0)\ R_y(0)} \qquad\qquad (5.6a)$$

For the cross-correlation of a random function $x(t)$ and its derivative $\dot{x}(t)$ one finds immediately

$$R_{x\dot{x}}(t,s) = \frac{\partial}{\partial s}\ E\left\{x(t)\ x(s)\right\} = \frac{\partial R_{xx}(t,s)}{\partial s} \qquad\qquad (5.7)$$

and, in the stationary case,

$$R_{x\dot{x}}(\tau) = \frac{dR_{xx}(\tau)}{d\tau} \qquad (5.7a)$$

If, in addition, x(t) is real, we have

$$R_{x\dot{x}}(0) = 0 \qquad (5.8)$$

as a consequence of the symmetry of R_{xx} with respect to $\tau = 0$.

Similarly, for the second derivative

$$R_{x\ddot{x}}(t,s) = \frac{\partial^2 R_{xx}(t,s)}{\partial s^2} \qquad \text{or} \qquad R_{x\ddot{x}}(\tau) = \frac{d^2 R_{xx}(\tau)}{d\tau^2} \qquad (5.9)$$

Stationary Processes.

Within the framework of correlation theory there is, of course, no difference between stationary and weakly stationary processes.

Ergodic Theorem

We are concerned here with the problem of finding the statistics of a process $x(t,\zeta)$ from a single observation $\zeta = \zeta_i$. If this is possible for the entire statistics the process is called ergodic in the most general form. For such a process time averages are equal to ensemble averages (expectations). Clearly, in correlation theory we are only interested in m and R(.).

Our first question is, therefore: under what circumstances is the time average

$$\bar{x} = \frac{1}{T} \int_0^T x(t) \, dt$$

in the limit $T \to \infty$ equal to the expectation $m = <x(t)>$? Obviously, $\bar{x}$ is a random quantity with expectation

$$E\left\{\bar{x}(t)\right\} = \frac{1}{T} \int_0^T m \, dt = m$$

The variance $\bar{\sigma}^2$ of $\bar{x}$ is given by

$$\bar{\sigma}^2 = E\left\{(\bar{x}-m)^2\right\} = \frac{1}{T^2} \int_0^T dt \int_0^T E\left\{\left[x(t)-m\right]\left[x(s)-m\right]\right\} ds =$$

$$= \frac{1}{T^2} \int_0^T dt \int_0^T \left[R(s-t) - m^2\right] ds$$

We introduce now the new variable $s - t = \tau$

$$ds = d\tau$$

In order to sweep the entire square area τ has to
go from $-T$ to $+T$. Thus we
have, writing $R(\tau)-m^2=A(\tau)$
for brevity,

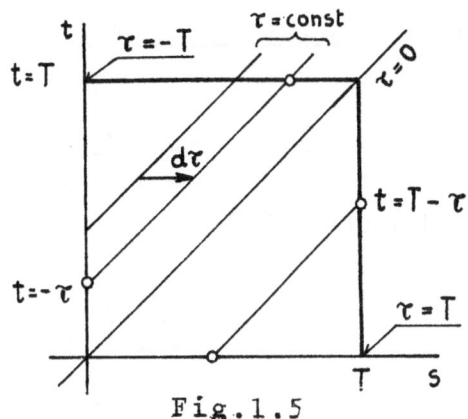

Fig.1.5

$$\bar{\sigma}^2 = \frac{1}{T^2}\left[\int_{-T}^{0}A(\tau)\int_{-\tau}^{T}dt\,d\tau + \int_{0}^{T}A(\tau)\int_{0}^{T-\tau}dt\,d\tau\right] =$$

$$= \frac{1}{T^2}\left[\int_{-T}^{0}(T+\tau)A(\tau)\,d\tau + \int_{0}^{T}(T-\tau)A(\tau)\,d\tau\right] =$$

$$= \frac{2}{T^2}\int_{0}^{T}(T-\tau)A(\tau)\,d\tau = \frac{2}{T}\int_{0}^{T}\left(1-\frac{\tau}{T}\right)\left[R(\tau)-m^2\right]d\tau$$

Now, in order to have the random variable $\bar{x}$ take on
its expected value m in the limit as $T\rightarrow\infty$ it is
necessary and sufficient that its variance $\bar{\sigma}^2$ goes
to zero in the limit. Since

$$\lim_{T\to\infty}\frac{2}{T}\int_{0}^{T}\left(1-\frac{\tau}{T}\right)\left[R(\tau)-m^2\right]d\tau = \lim_{T\to\infty}\frac{2}{T}\int_{0}^{T}\left[R(\tau)-m^2\right]d\tau$$

we have <u>ergodic theorem I :</u>

$$\lim_{T \to \infty} \frac{1}{T} \int_0^T x(t)\, dt = m \qquad (5.10)$$

if and only if

$$\lim_{T \to \infty} \frac{1}{T} \int_0^T R(t)\, dt = m^2 \qquad (5.11)$$

If we now consider the process

$$u(t) = x(t+\tau)\, x(t)$$

whose expectation is $R(\tau)$ and whose autocorrelation is given by

$$R_{uu}(\lambda) = E\left\{ x(t+\tau)\, x(t)\, x(t+\tau+\lambda)\, x(t+\lambda) \right\} \qquad (5.12)$$

we obtain at once <u>ergodic theorem II :</u>

$$\lim_{T \to \infty} \frac{1}{T} \int_0^T x(t+\tau)\, x(t)\, dt = R(\tau) \qquad (5.13)$$

if and only if

$$\lim_{T \to \infty} \frac{1}{T} \int_0^T R_{uu}(\lambda)\, d\lambda = R^2(\tau) \qquad (5.14)$$

We observe that a knowledge of m and $R(.)$ suffices to test a process for ergodicity of the mean. To test for ergodicity of the autocorrelation one has to know the fourth-order moment R_{uu} !

Spectral Density.

The   Fourier transform

$$X(\omega) = \int_{-\infty}^{+\infty} x(t) e^{-i\omega t} dt \quad , \quad x(t) = \frac{1}{2\pi} \int_{-\infty}^{+\infty} X(\omega) e^{i\omega t} d\omega$$

$$(5.15)$$

does, in general, not exist for a stationary stochastic process. On the other hand, the autocorrelation does have a Fourier transform called spectral density S (ω):

$$S(\omega) = \int_{-\infty}^{+\infty} R(t) e^{-i\omega t} dt = 2 \int_{0}^{\infty} R(t) \cos \omega t \, dt \qquad (5.16)$$

$$R(t) = \frac{1}{2\pi} \int_{-\infty}^{+\infty} S(\omega) e^{i\omega t} d\omega = \frac{1}{\pi} \int_{0}^{\infty} S(\omega) \cos \omega t \, d\omega \qquad (5.17)$$

The two Eqs. (5.16/17) are known as the Wiener-Chintchine equations [*]. It follows from (5.16) that S(ω) is real and even.

If ergodic theorem II , Eq. (5.13), is assumed to hold (equality of sample correlation and ensemble correlation) Eq. (5.16) may be written

---

[*] Watch for different definitions of the Fourier transform in different sources!

$$S(\omega) = \lim_{T \to \infty} \frac{1}{2T} \int_{-\infty}^{+\infty} e^{-i\omega\tau} \int_{-T}^{T} x(t+\tau)\, x(t)\, dt\, d\tau \qquad (5.18)$$

One also arrives at this expression in the following manner. Replacing the divergent Fourier-transform (5.15) by

$$X(\omega) = \lim_{T \to \infty} \frac{1}{\sqrt{2T}} \int_{-T}^{+T} x(t)\, e^{-i\omega t}\, dt \qquad (a)$$

one gets

$$X(\omega) X^*(\omega) = \lim_{T \to \infty} \frac{1}{2T} \int_{-T}^{T} x(\lambda) \int_{-T}^{+T} x(t)\, e^{-i\omega(\lambda-t)}\, dt\, d\lambda$$

which, putting

$$\lambda - t = \tau \quad , \quad d\lambda = d\tau$$

gives at once

$$S(\omega) = X(\omega)\, X^*(\omega) \geq 0 \qquad (5.19)$$

Transformation (a) is called "generalized Fourier transform".

Linear Systems

The output of a linear system with con̲stant coefficients may be represented in the form (4.8)[*]

---

[*] Using the δ-function and its derivatives, if necessary.

$$y(t) = \int_{-\infty}^{t} h(t-\tau)x(\tau)\,d\tau = \int_{0}^{\infty} h(\tau)x(t-\tau)\,d\tau \qquad (5.20)$$

If the process started at $t=0$, then $x(t)=0$ for $t<0$, and (5.20) goes over into

$$y(t) = \int_{0}^{t} h(t-\tau)x(\tau)\,d\tau = \int_{0}^{t} h(\tau)x(t-\tau)\,d\tau \qquad (5.21)$$

The function $h(t)$ is called "impulse response" of the system since it represents the response to $x(t) = \delta(t)$ as may be seen from (5.20). By "transition function"[*] $U(t)$ is meant the response of the system to a unit step input $x(t) = H(t)$:

$$U(t) = \int_{0}^{t} h(t-\tau)\,d\tau = \int_{0}^{t} h(\lambda)\,d\lambda \qquad (5.22)$$

Finally, if the input to the system is periodic with unit amplitude, $x(t) = e^{i\omega t}$, one obtains, after the transient response has died out, a purely periodic output whose complex amplitude $F(\omega)$ is known as the "frequency response" or "admittance" of the system:

$$y(t) = F(\omega)e^{i\omega t} = \int_{-\infty}^{t} h(t-\tau)e^{i\omega\tau}\,d\tau$$

One finds

$$F(\omega) = \int_{-\infty}^{t} h(t-\tau)e^{-i\omega(t-\tau)}\,d\tau = \int_{0}^{\infty} h(\lambda)e^{-i\omega\lambda}\,d\lambda \qquad (5.23)$$

[*] not to be confused with "transfer function".

Hence, $F(0) = U(\infty)$.

If in particular, the system may be described by a linear differential equation with constant coefficients,

$$P(D) y(t) = x(t) H(t) \qquad (5.24)$$

with $P(D)$ given by (4.7), then $h(t)$ is that solution of the homogeneous differential equation (5.24)

$$P(D) h(t) = 0 \qquad (5.25)$$

which satisfies the initial condition

$$h(0) = h'(0) = \ldots = h^{(n-2)}(0) = 0 \ , \ h^{(n-1)}(0) = \frac{1}{a_n} \qquad (5.26)$$

Putting $x = e^{i\omega t}$, $y = F(\omega) e^{i\omega t}$ in (5.24), one obtains for the admittance

$$F(\omega) = \frac{1}{P(i\omega)} \qquad (5.27)$$

If, instead of (5.24) one has the differential equation

$$P(D) y(t) = Q(D) x(t) \qquad (5.28)$$

then

$$F(\omega) = \frac{Q(i\omega)}{P(i\omega)} \qquad . \qquad (5.29)$$

We consider now the case where $x(t)$ is a stationary random function. The expectation m_y

is then,

$$m_y = \int_0^t h(t-\tau)\, m_x\, d\tau = m_x\, U(t) \qquad (5.30)$$

After a sufficiently long time this becomes

$$m_y = F(0)\, m_x = \text{const.} \qquad (5.31)$$

Eq.(4.6) gives for the autocorrelation of the output y

$$R_{yy}(t,s) = \int_0^t h(t-\tau) \int_0^s h(s-\lambda)\, R_{xx}(\lambda-\tau)\, d\lambda\, d\tau$$

$$(5.32)$$

Integration is, in general, difficult. However, upon introduction of the spectral density with the aid of (5.17)

$$R_{xx}(\lambda-\tau) = \frac{1}{\pi} \int_0^\infty S_{xx}(\omega)\, \cos\omega(\lambda-\tau)\, d\omega$$

Eq. (5.32) takes the form

$$R_{yy}(t,s) = \frac{1}{\pi} \int_0^\infty S_{xx}(\omega)\left\{ \int_0^t \left[h(t-\tau) \int_0^s h(s-\lambda)\cos\omega(\lambda-\tau)\, d\lambda \right] d\tau \right\} d\omega$$

$$(5.33)$$

which, sometimes, is more convenient for integration. One observes that during the transient period $y(t)$ is not a stationary random function since R_{yy} then depends on both t and s, and not on the difference $t-s$. Only after a sufficiently long time, $t \rightarrow \infty$, $y(t)$ becomes stationary.

Some fundamental relations for this stationary case will now be derived. Multiplying both sides of the second form of (5.20) by $x(t-\lambda)$, and taking expectations, we obtain

$$E\left\{y(t)x(t-\lambda)\right\} = R_{yx}(\lambda) = \int_0^\infty h(\tau)E\left\{x(t-\tau)x(t-\lambda)\right\}d\tau = \int_0^\infty h(\tau)R_{xx}(\lambda-\tau)d\tau$$

$$(5.34)$$

We multiply now both sides of this equation by $e^{-i\omega\lambda}$ and integrate between $-\infty$ and $+\infty$. We get

$$\int_{-\infty}^{+\infty} R_{yx}(\lambda)e^{-i\omega\lambda}d\lambda = S_{yx}(\omega) = \int_0^\infty h(\tau)\int_{-\infty}^\infty R_{xx}(\lambda-\tau)e^{-i\omega\lambda}d\lambda\,d\tau =$$

$$= \int_0^\infty h(\tau)e^{-i\omega\tau}d\tau \int_{-\infty}^{+\infty} R_{xx}(\varepsilon)e^{-i\omega\varepsilon}d\varepsilon$$

or, taking (5.23) into account,

$$S_{yx}(\omega) = S_{xx}(\omega)F(\omega) \qquad\qquad (5.35)$$

Now we repeat the procedure by multiplying (5.20) with $y(t+\lambda)$. Taking expectations we find

$$E\left\{y(t)y(t+\lambda)\right\} = R_{yy}(\lambda) = \int_0^\infty h(\tau)E\left\{x(t-\tau)\cdot y(t+\lambda)\right\}d\tau = \int_0^\infty h(\tau)R_{yx}(\lambda+\tau)d\tau$$

$$(5.36)$$

If, again, this is multiplied by $e^{-i\omega\lambda}$ and integrated, one has

$$S_{yy}(\omega) = \int_{-\infty}^{+\infty} R_{yy}(\lambda)e^{-i\omega\lambda}d\lambda = \int_{0}^{\infty} h(\tau) \int_{-\infty}^{\infty} R_{yx}(\lambda+\tau)e^{-i\omega\lambda}d\lambda\, d\tau =$$

$$= \int_{0}^{\infty} h(\tau)e^{+i\omega\tau}d\tau \int_{-\infty}^{+\infty} R_{yx}(\varepsilon)e^{-i\omega\varepsilon}d\varepsilon$$

and, therefore,

$$S_{yy}(\omega) = S_{yx}(\omega)\, F^{*}(\omega) \qquad\qquad (5.37)$$

Finally, upon substituting (5.35) into (5.37), one obtains the following basic relation between the spectral densities of input and output of a linear system in the stationary state

$$S_{yy}(\omega) = F(\omega)F^{*}(\omega)\, S_{xx}(\omega) \qquad\qquad (5.38)$$

$F(\omega)\, F^{*}(\omega)$ is the square of the absolute value of the admittance and is, therefore, a real quantity.

As an example, the spectral density $S_{\dot{x}\dot{x}}$ of the derivative $y = dx/dt$ of a stationary process $x(t)$ is, with $F(\omega) = i\omega$,

$$S_{\dot{x}\dot{x}}(\omega) = \omega^2 S_{xx}(\omega) \qquad\qquad (5.39a)$$

Similarly

$$S_{\ddot{x}\ddot{x}}(\omega) = \omega^4 S_{xx}(\omega) \qquad\qquad (5.39b)$$

Some Important Correlation Functions and Spectral Densities

$x(t)$	$R(\tau)$	$S(\omega)$
$ax(t)$	$aa^* R(\tau)$	$aa^* S(\omega)$
$\dot{x}(t)$	$-\dfrac{d^2 R}{d\tau^2}$	$\omega^2 S(\omega)$
$\dfrac{d^n x}{dt^n}$	$(-1)^n \dfrac{d^{2n} R}{d\tau^{2n}}$	$\omega^{2n} S(\omega)$
$x(t)e^{\pm i\omega_0 t}$	$R(\tau)e^{\pm i\omega_0 \tau}$	$S(\omega \pm \omega_0)$
(1)	$e^{-\alpha|\tau|}$	$\dfrac{2\alpha}{\omega^2 + \alpha^2}$
(2)		$\dfrac{4}{T\omega^2}\sin^2\left(\dfrac{\omega T}{2}\right)$
(3)	1	$2\pi\delta(\omega)$
(4) band-limited noise	$\dfrac{2}{\pi}\dfrac{\sin\alpha\tau}{\tau}$	
(5) white noise	$\delta(\tau)$	1

(6)	$\cos \omega_0 \tau$	$\pi\left[\delta(\omega-\omega_0)+\delta(\omega+\omega_0)\right]$
(7)	$e^{-\alpha|\tau|}\cos\beta\tau$	$\dfrac{2\alpha(\omega^2+\alpha^2+\beta^2)}{(\omega^2-\beta^2-\alpha^2)^2+4\alpha^2\omega^2}$
(8)	$e^{-\alpha|\tau|}\left(\cos\beta\tau+\dfrac{\alpha}{\beta}\sin\beta|\tau|\right)$	$\dfrac{4\alpha(\alpha^2+\beta^2)}{(\omega^2-\beta^2-\alpha^2)^2+4\alpha^2\omega^2}$
(9)	$e^{-\alpha^2\tau^2}\cos\beta\tau$	$\dfrac{\sqrt{\pi}}{2\alpha}\left[\exp\left(-\dfrac{(\omega+\beta)^2}{4\alpha^2}\right)+\exp\left(-\dfrac{(\omega-\beta)^2}{4\alpha^2}\right)\right]$
(10)	$\displaystyle\sum_{k=0}^{n} a_k\, \delta^{(k)}(\tau)$	$\displaystyle\sum_{k=0}^{n}(i\omega)^k a_k\,,\quad \left[\delta^{(k)}=\dfrac{d^k}{dt^k}\,\delta(t)\right]$

N o t e.[*)]   We have shown (p. 32) that for the
existence of the derivative in mean square $\dot{x}(t)$ of a
random process $x(t)$ the existence of the first and
second derivatives in $\tau = 0$ of the autocorrelation is
necessary and sufficient. Now, in general, there can
be no doubt as to the differentiability of the random
functions occurring in engineering applications. However, the analytical expressions chosen to approximate the correlation functions obtained in experiment
are frequently so simple, cf. (1), (7), (8) in the
preceding table, that they belong to nondifferentiable
processes like (1) and (7), or to processes that are
only differentiable once, like (8).

        It is, therefore, of great importance
that, at least in the stationary case, it can be
shown that correlation theory and, in particular, the
fundamental relation (5.38) remain valid even if,
formally, the derivatives of arbitrary order of nondifferentiable functions are involved.

        Consider differential equation (5.28).
It suffices to take $Q(D) = D^m$. Eq. (5.20) then
renders for the stationary solution

$$y(t) = \int_0^\infty h(\tau) \ \frac{d^m x(t-\tau)}{d\tau^m} \ d\tau$$

[*)] See $\begin{bmatrix} 8 \end{bmatrix}$, p. 80

Repeated integration by parts transforms this equation into a form which no longer contains the derivatives of $x(t)$. The integrated parts drop out as a
consequence of (5.26) and we obtain

$$y(t) = (-1)^m \int_0^\infty \frac{d^m h(\tau)}{d\tau^m} \, x(t-\tau) \, d\tau$$

For the correlation function of $y(t)$ one has, therefore,

$$R_{yy}(\lambda) = E\left\{ y(t) \, y(t+\lambda) \right\} = \int_0^\infty \int_0^\infty \frac{d^m h(\tau)}{d\tau^m} \frac{d^m h(s)}{ds^m} R_{xx}(\lambda + \tau - s) \, d\tau \, ds$$

Employing (5.17) to express R_{xx} in terms of the spectral density $S_{xx}(\omega)$ we get

$$R_{yy}(\lambda) = \frac{1}{2\pi} \int_{-\infty}^{+\infty} e^{i\omega\lambda} \left[\int_0^\infty \frac{d^m h(\tau)}{d\tau^m} e^{i\omega\tau} \, d\tau \int_0^\infty \frac{d^m h(s)}{ds^m} e^{-i\omega s} \, ds \right] S_{xx}(\omega) \, d\omega$$

If we now again integrate by parts the integrated
parts vanish and we obtain

$$R_{yy}(\lambda) = \frac{1}{2\pi} \int_{-\infty}^{+\infty} e^{i\omega\lambda} \left| (i\omega)^m \right|^2 \left| \int_0^\infty h(\tau) e^{-i\omega\tau} \, d\tau \right|^2 S_{xx}(\omega) \, d\omega$$

According to (5.23) the second integral represents
the admittance of the left side of (5.28), i.e.,
$1/P(i\omega)$, while $(i\omega)^m$ is identical with $Q(i\omega)$.
Using (5.29) we have, therefore, obtained relation
(5.38) without assuming the existence of the derivative $d^m x / dt^m$!

6. The Poisson Process .

As a first special random process of considerable practical interest we discuss the Poisson process.

Let an event occur in irregular time intervals. Examples are the emission of an electron from the cathode of a tube, or the arrival of customers at a service station, or, also, the occurrence of accidents, the rupture of a machine component, etc.

Common to all these events is that they may be characterized by a <u>counting function</u> $x(t)$ giving the number of events that occurred in the time interval $(0,t]$, open on the left, provided we start counting at time $t = 0$. Hence, the random variable x can take on only the integer values $0,1,2.......$, and $x(t)$ is a discontinuous process. Initially $x(0) = 0$. Such a counting process is called Poisson process provided it satisfies the following conditions:

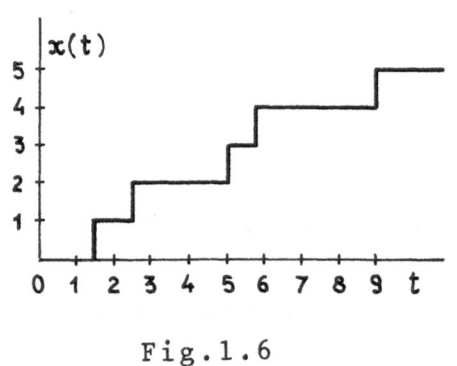

Fig.1.6

(I) The <u>increments</u> $x(t)-x(s)$ are <u>independent</u>. If the random variables $x(t)-x(s)$ and $x(t+h)-x(s+h)$ have the same p.-distribution we speak of a process with <u>stationary increments</u>.

(II) $P\left[x(t+s) - x(t) = 1\right] = \nu s$ (6.1)

              $P\left[x(t+s) - x(t) \geqslant 2\right] = 0$ for $s \to 0$

              $P\left[x(t+s) - x(t) = 0\right] = 1 - \nu s$

In words: In a sufficiently small time-interval s the probability that <u>one</u> event will occur is proportional to s, while the probability that <u>two or more</u> events will occur is zero. Hence, the probability that <u>either one or no event at all</u> occurs is one.

The factor of proportionality ν may be constant (homogeneous process), or may depend on time, $\nu = \nu(t)$, (inhomogeneous process).

We look first for the probabilities

$$P\left[x(t) = n\right] \equiv p_n(t) \qquad (n = 0, 1, 2 \ldots) \qquad (6.2)$$

At two arbitrary time instants t and t+s we have from condition (I)

$$p_n(t+s) = p_0(t)q_n(s) + p_1(t)q_{n-1}(s) + \ldots p_n(t)q_0(s) = \sum_{i=0}^{n} p_i(t)q_{n-i}(s)$$

(6.3)

We expand in powers of s and let $s \to 0$. Using the initial condition

$$p_n(0) = 0 \qquad (6.4)$$

and condition (II) for the conditional probabilities

$$q_n(s) = 0 \quad \text{for } n \geq 2$$

$$q_1(s) = \nu s \qquad \qquad \Bigg\} \quad s \longrightarrow 0 \qquad (6.5)$$

$$q_0(s) = 1 - \nu s$$

we find

$$p_n(t) + s \; \frac{dp_n(t)}{dt} + \ldots = \ldots \; p_{n-1}(t) \nu s + p_n(t) \left[1 - \nu s \right]$$

or

$$\frac{dp_n(t)}{dt} + \nu p_n(t) = \nu p_{n-1}(t) \qquad (6.6)$$

The solution of this linear differential-difference-equation is, for ν =const, taking initial condition (6.4) into consideration, given by

$$p_n(t) = e^{-\nu t} \; \frac{(\nu t)^n}{n!} \qquad (6.7)$$

The parameter ν is called density or intensity of the Poisson process. If ν is not constant the Poisson process is called non-homogeneous and can be reduced to a homogeneous process. In the following ν will be assumed to be constant.

Expectation $E\{x(t)\} = m(t)$ is, according to (3.8), given by

$$m(t) = \sum_{n=1}^{\infty} n\, p_n(t) = e^{-\nu t} \sum_{n=1}^{\infty} n \frac{(\nu t)^n}{n!}$$

Upon differentiating the series

$$e^x = \sum_{n=0}^{\infty} \frac{x^n}{n!}$$

termwise with respect to x one obtains

$$e^x = \sum_{n=1}^{\infty} n \frac{x^{n-1}}{n!} = \frac{1}{x} \sum n \frac{x^n}{n!}$$

and, hence,

$$m(t) = \nu t \qquad\qquad (6.8)$$

In the same manner one finds for

$$E\{x^2(t)\} = \sum_{n=1}^{\infty} n^2\, p_n(t) = e^{-\nu t} \sum_{n=1}^{\infty} n^2 \frac{(\nu t)^n}{n!}$$

by differentiating the series for e^x twice

$$E\{x^2(t)\} = (\nu t)^2 + \nu t$$

The variance is, therefore,

$$\sigma_x^2 = \nu t \qquad (6.9)$$

For the random variable $x(t) - x(s)$ we find, using (I) and Eq.(2.28),

$$P\left[x(t) - x(s) = n\right] = p_n(t-s) \qquad (6.10)$$

Hence, the increments are also stationary, and

$$E\left[x(t) - x(s)\right] = \nu(t-s) \qquad (6.11)$$

$$E\left\{\left[x(t) - x(s)\right]^2\right\} = \nu^2(t-s)^2 + \nu(t-s) \qquad (6.12)$$

We consider now four time instances t_1, t_2, t_3, t_4. If the corresponding intervals do not overlap, $t_1 > t_2 > t_3 > t_4$, the variables $x(t_1) - x(t_2)$ and $x(t_3) - x(t_4)$ are independent and we have

$$E\left\{\left[x(t_1) - x(t_2)\right]\left[x(t_3) - x(t_4)\right]\right\} = E\left[x(t_1) - x(t_2)\right] E\left[x(t_3) - x(t_4)\right] =$$

$$= \nu^2(t_1 - t_2)(t_3 - t_4) \qquad (6.13)$$

If, on the other hand, the intervals do overlap, $t_1 > t_3 > t_2 > t_4$, Eq. (6.13) no longer holds.

Putting, however,

$$x(t_1) - x(t_2) = \left[x(t_1) - x(t_3)\right] + \left[x(t_3) - x(t_2)\right]$$

$$x(t_3) - x(t_4) = \left[x(t_3) - x(t_2)\right] + \left[x(t_2) - x(t_4)\right]$$

Eq. (6.13) can be used again and one obtains after some algebra, using (6.12)

$$E\left\{\left[x(t_1) - x(t_2)\right]\left[x(t_3) - x(t_4)\right]\right\} = \nu^2(t_1 - t_2)(t_3 - t_4) + \nu(t_3 - t_2) \qquad (6.14)$$

$t_3 - t_2$ is the length of the overlapping interval. Putting $t_1 = t$, $t_3 = s$, $t_2 = t_4 = 0$ in Eq. (6.14) the _autocorrelation_ follows as

$$R(t,s) = E\left\{x(t)x(s)\right\} = \nu^2 ts + \nu s \qquad (t \geq s) \qquad (6.15)$$

The characteristic function is

$$\varphi(q) = \exp\left[\nu t\left(e^{iq} - 1\right)\right] \qquad (6.16)$$

7. The Wiener Process .

This process, named after Norbert Wiener, represents the mathematical model of the Brownian mo tion. However, it seems to be applicable to entirely different phenomena too, the stock market, for instance $[10]$.

We consider a particle in a medium in equilibrium. Since it is continuously being pushed by the fluid or gas molecules it describes a random path. Its displacement $x(t) - x(s)$, in a time interval which is large in comparison with the time between two subsequent pushes, may be considered as the sum of a large number of small irregular displacements. According to the central limit theorem it appears, therefore, reasonable to consider $x(t) - x(s)$ as being normally distributed and, since the surrounding medium is in equilibrium, as stationary[*]. Finally, we assume that the displacements or displacement increments over non-overlapping time intervals are independent.

This leads to the following definition of the Wiener process:

(I) $\{x(t), t \geq 0\}$ has independent stationary increments.

(II) For each $t > 0$ $x(t)$ is normally distributed.

(III) For all $t > 0$ $E\{x(t)\} = 0$.

(IV) $x(0) = 0$

---

[*] The process $x(t)$ itself is not stationary!

From (II) and (III) we have at once for the p.-density

$$p(x;t) = \frac{1}{\sqrt{2\pi}\;\sigma(t)}\exp\left[-\frac{x^2}{2\sigma^2(t)}\right] \qquad (7.1)$$

Furthermore, using assumption (IV),

$$\text{Var}\left\{x(t+\tau)\right\} = \text{Var}\left\{\left[x(t+\tau)-x(t)\right] + \left[x(t)-x(0)\right]\right\} =$$

$$= E\left\{\left[x(t+\tau)-x(t)\right]^2\right\} + E\left\{\left[x(t)-x(0)\right]^2\right\} + 2E\left\{\left[x(t+\tau)-x(t)\right]\left[x(t)-x(0)\right]\right\}$$

Since, by assumption (I), the increments are independent the last term is equal to the product of the expectations and vanishes therefore as a consequence of assumption (III). Noting finally that, by assumption (I), the increments are also stationary,

$$E\left\{\left[x(t+\tau)-x(t)\right]^2\right\} = E\left\{\left[x(\tau)-x(0)\right]^2\right\} = E\left\{\left[x(\tau)\right]^2\right\}$$

one obtains the following functional equation for the variance of the Wiener process

$$\text{Var}\left\{x(t+\tau)\right\} = \text{Var}\left\{x(t)\right\} + \text{Var}\left\{x(\tau)\right\}$$

The only continuous, nonnegative solution of this equation is

$$\text{Var}\left\{x(t)\right\} = \sigma^2(t) = \alpha t \qquad (7.2)$$

Hence, also

$$\text{Var}\left\{x(t) - x(s)\right\} = \alpha\,|t-s| \qquad (7.3)$$

Eq. (7.1) now goes over into

$$p(x;t) = \frac{1}{\sqrt{2\pi\alpha t}}\,\exp\left[-\frac{x^2}{2\alpha t}\right] \qquad (7.4)$$

The constant α follows from experiment. For the Brownian motion it was also determined theoretically by Einstein, see [11] , p.93.

The Wiener process may be considered as the integral of Gaussian <u>white noise</u>:

$$x(t) = \int_0^t w(\tau)\,d\tau \qquad (7.5)$$

Indeed, all assumptions (I) to (IV) are satisfied. Normality follows from the linearity of the integral operator while for the variance one has, using (4.6) with $R_w(\tau) = \alpha\delta(\tau)$

$$\sigma^2(t) = \alpha\int_0^t ds \int_0^t \delta(\tau)\,d\tau = \alpha\int_0^t ds = \alpha t$$

Last we show that the conditional p.-density $p(x;t\,|\,a\,;s)$ of the Wiener process satisfies the diffusion equation

$$\frac{\partial p}{\partial t} = \frac{\alpha}{2} \frac{\partial^2 p}{\partial x^2} \qquad (7.6)$$

Since $x(t)$ is normal with zero mean we have for the conditional expectation from Eq. (2.38)

$$E\left\{x(t)\,\big|\,x(s)=a\right\} = \frac{E\left\{x(t)\,x(s)\right\}}{E\left\{x^2(s)\right\}}\cdot a$$

However, from Eq. (7.3) for $t \geq s$,

$$E\left\{[x(t)-x(s)]^2\right\} = E\left\{[x(t)]^2\right\} + E\left\{[x(s)]^2\right\} - 2E\left\{x(t)\,x(s)\right\} = \alpha\,(t-s)$$

whence, using Eq. (7.2),

$$E\left\{x(t)\,x(s)\right\} = \alpha s \qquad (t \geq s) \qquad (7.7)$$

and

$$E\left\{x(t)\,\big|\,x(s)=a\right\} = a \qquad (7.8)$$

Similarly, one finds from Eq. (2.39) for the conditional variance

$$\text{Var}\left\{x(t)\,\big|\,x(s)=a\right\} = E\left\{[x(t)-a]^2\,\big|\,x(s)=a\right\} = \alpha t - \frac{(\alpha s)^2}{\alpha s} = \alpha(t-s)$$

$$(7.9)$$

The conditional p.-density of the Wiener process is, therefore,

$$p(x;t\,|\,a;s) = \frac{1}{\sqrt{2\pi\alpha(t-s)}}\; \exp\left[-\frac{(x-a)^2}{2\alpha(t-s)}\right] \qquad (7.10)$$

which may easily be shown to satisfy Eq. (7.6).

Eq. (7.6) is a special form of the Fokker-Planck equation to be discussed in the next section.

8. Markov Sequences .

As a preliminary to the discussion of the Markov process we first briefly discuss Markov sequences.

Through Eq.(1.4) we have introduced con-ditional probability. Consider now a sequence $x_1, x_2 \ldots x_n \ldots$ of random variables[*]. The expres-sion

$$P\left[x_n \leq X_n\,|\,x_{n-1}=X_{n-1},\,x_{n-2}=X_{n-2}\ldots x_1=X_1\right] = F\left(X_n\,|\,X_{n-1},\ldots X_1\right)$$

$$(8.1)$$

---

[*] Such a sequence may be looked upon as a special case of a random process $x(t)$ where only the values $x_i = x(t_i)$ at certain discrete time instants $t_1, t_2 \ldots$ are regis-tered.

then represents the probability that $x_n \leq X_n$
if one already knows the values $X_{n-1}, X_{n-2} \ldots X_1$
that have been taken on by the preceding terms $x_{n-1}, x_{n-2} \ldots x_1$ in the sequence.

In general, each preceding value will have some influence on the probability of x_n. If, however, only the value X_{n-1} immediately preceding x_n influences the probability of x_n while all other values $X_{n-2} \ldots X_1$ are without influence,

$$F\left(X_n \mid X_{n-1}, X_{n-2} \ldots X_1\right) = F\left(X_n \mid X_{n-1}\right) \qquad (8.2)$$

we speak of a Markov sequence. The conditional probability (8.2) is frequently called transition probability.

For a continuous sequence a p.-density exists given by

$$p\left(x_n \mid x_{n-1}, x_{n-2} \ldots x_1\right) = p\left(x_n \mid x_{n-1}\right) \qquad (8.3)$$

Therefore, while the joint p.-density for an arbitrary sequence is, according to Eq. (1.4), given by

$$p\left(x_1, x_2 \ldots x_n\right) = p\left(x_1\right) p\left(x_2 \mid x_1\right) p\left(x_3 \mid x_2, x_1\right) \ldots (8.4)$$

this reduces for a Markov sequence to

$$p\left(x_1, x_2 \ldots x_n\right) = p\left(x_n \mid x_{n-1}\right) p\left(x_{n-1} \mid x_{n-2}\right) \ldots p\left(x_2 \mid x_1\right) p\left(x_1\right) \qquad (8.5)$$

From (8.3) we conclude that

$$E\left\{x_n | x_{n-1}, x_{n-2} \ldots x_1\right\} = E\left\{x_n | x_{n-1}\right\} \qquad (8.6)$$

The two random variables x_n and x_s of a Markov sequence are, for $n > r > s$ conditionally independent, since

$$p(x_n, x_s | x_r) = p(x_n | x_r) \, p(x_s | x_r) \qquad (8.7)$$

Proof: Since (8.5) holds for any subsequence of a Markov sequence it follows from (1.4) that

$$p(x_n, x_s | x_r) = \frac{p(x_n, x_r, x_s)}{p(x_r)} = \frac{p(x_n | x_r) p(x_r | x_s) p(x_s)}{p(x_r)}$$

whence, because of $p(x_r | x_s) \, p(x_s) = p(x_r) \, p(x_s | x_r)$ [*], Eq. (8.7) follows.

Integration over a random variable from $-\infty$ to $+\infty$ eliminates it from the p.-density:

$$p(x_n, x_s) = \int_{-\infty}^{+\infty} p(x_n, x_r, x_s) \, dx_r \qquad (8.8)$$

---

[*] A Markov sequence is Markov also in reverse:

$$p(x_n | x_{n+1}, x_{n+2} \ldots x_{n+k}) = p(x_n | x_{n+1})$$

Similarly, for the conditional density,

$$p\left(x_n \mid x_s\right) = \int\limits_{-\infty}^{+\infty} p\left(x_n \mid x_r, x_s\right) p\left(x_r \mid x_s\right) dx_r \qquad (8.9)$$

If this rule for the elimination of a random variable is applied to a Markov sequence the Chapman - Kolmogorov integral equation results:

$$p\left(x_n \mid x_s\right) = \int\limits_{-\infty}^{+\infty} p\left(x_n \mid x_r\right) p\left(x_r \mid x_s\right) dx_r \qquad (8.10)$$

or, upon multiplication by $p\left(x_s\right)$ and integration

$$p\left(x_n\right) = \int\limits_{-\infty}^{+\infty} p\left(x_n \mid x_r\right) p\left(x_r\right) dx_r \qquad (8.11)$$

If the Markov sequence x_i can only take on the discrete values

$$a_1, a_2 \ldots a_N$$

such that

$$P\left[x_n = a_i \mid x_{n-1} = a_j, \; x_{n-2} = a_k, \ldots x_1 = a_l\right] = P\left[x_n = a_i \mid x_{n-1} = a_j\right] \quad (8.12)$$

it is called a Markov chain. Using the following notation for absolute and conditional probabilities

$$P\left[x_n = a_i\right] = p_i(n), \quad P\left[x_n = a_i \mid x_s = a_j\right] = p_{ij}(n,s) \qquad (8.13)$$

one immediately has the relations

$$\sum_{i=1}^{N} p_i(n) = 1 \; , \; \sum_{i=1}^{N} p_{ij}(n,s) = 1 \qquad (8.14)$$

$$p_{ij}(n,s) = \sum_{k=1}^{N} p_{ik}(n,r) \, p_{kj}(r,s) \qquad (8.15)$$

$$p_i(n) = \sum_{j=1}^{N} p_{ij}(n,s) \, p_j(s) \qquad (8.16)$$

The last two equations are the discrete version of Eq. (8.10) and (8.11).

If, for every j , the "expectation" of the p.-distribution p_{jk} equals a_j ,

$$\sum_{j} a_j p_{jk} = a_k \qquad (8.17)$$

the Markov chain is called a <u>martingale</u>[*]

9. Markov Processes .

In consequent extension of Eq. (8.2) a random process is called Markov process if

$$P\left[x(t_n) \leqslant X_n \,\middle|\, x(t_{n-1}) = X_{n-1}, \dots x(t_1) = X_1\right] = P\left[x(t_n) \leqslant X_n \,\middle|\, x(t_{n-1}) = X_{n-1}\right]$$
$$(9.1)$$

[*] Word from the French Provence denoting systematic betting.

for every $t_n > t_{n-1} \ldots > t_1$. In other words: if $x(t_0)$ has been fixed no condition that may be imposed on $\{x(t), t < t_0\}$ has any influence upon the statistics for $t > t_0$. Statistical behavior in the future is uniquely determined by the present, independent of the past.

If, for a process, $x(t_2) - x(t_1)$ for $t_2 > t_1$ is independent of $x(t)$ for every $t < t_1$, the process is Markov. Indeed, $x(t_2)$ then depends only on $x(t_1)$ but not also on any preceding value. It follows, in particular, that a process with independent increments is Markov, provided $x(0) = 0$, since then $x(t_2) - x(t_1)$ is independent of $x(t) - x(0) = x(t)$ for all $t < t_1$. Hence, both the Poisson process and the Wiener process are Markov.

The equations of the preceding section remain valid provided x_n is replaced by $x(t_n)$. For instance, Eq. (8.5) for a continuous process becomes

$$p(x_1, x_2 \ldots x_n ; t_1, t_2 \ldots t_n) = p(x_1 ; t_1) \prod_{r=2}^{n} p(x_r ; t_r \,|\, x_{r-1} ; t_{r-1})$$

$$(9.2)$$

The <u>Chapman - Kolmogorov</u> equation (8.10) reads now

$$p(x ; t \,|\, x_0 ; t_0) = \int_{-\infty}^{+\infty} p(x ; t \,|\, x_1 ; t_1) \, p(x_1 ; t_1 \,|\, x_0 ; t_0) \, dx_1$$

$$(9.3)$$

and Eq. (8.11) becomes

$$p\left(x;t\right) = \int_{-\infty}^{+\infty} p\left(x;t \,|\, x_0;t_0\right) p\left(x_0;t_0\right) dx_0$$

$$(9.4)$$

Analogous relations hold for n-dimensional processes

$$x\left(t\right) = \left[x_1(t), \; x_2(t) \ldots x_n(t)\right]$$

If a process $x\left(t\right)$ is determined by the initial value problem

$$\frac{dx}{dt} = f\left(x;t\right) + g\left(t\right), \qquad x\left(t_0\right) = x_0 \qquad (9.5)$$

and if the random function $g(t)$ has the property that $g\left(t_1\right), g\left(t_2\right) \ldots g\left(t_n\right)$ are independent [*] for every $t_1 < t_2 \ldots\ldots\ldots < t_n$, then $x\left(t\right)$ is a Markov process. Indeed, $x\left(t\right)$ is uniquely determined by its initial value $x\left(t_0\right)$ and the values $g\left(t\right)$ in the interval $\left(t_0, t\right)$. All preceding values of $x\left(t\right)$ are of no influence.

---

[*] $g\left(t\right)$ is then called a <u>purely random function</u>, and is defined completely by its p.-density $p_1\left(x;t\right)$, since then

$$p_n\left(x_1 \ldots x_n; \; t_1 \ldots t_n\right) = p_1\left(x_1;t_1\right) \ldots p_1\left(x_n;t_n\right)$$

An extreme case is <u>white noise</u>.

The same holds for a differential equation of order n which can always be reduced to a system of n differential equations of order one and hence, to an n-dimensional Markov process.

For a <u>normally distributed</u> Markov process with mean value zero

$$R\left(t_2,t\right) R\left(t,t_1\right) = R\left(t_2,t_1\right) R\left(t,t\right) \qquad (9.6)$$

for all $t_2 > t > t_1$. <u>Proof</u>: Let $x(t_2) = u$, $x(t) = y$, $x(t_1) = z$ and construct a function $w = u - \alpha y - \beta z$, orthogonal to y and z cf.p.21. Then

$$E\left\{u - \alpha y - \beta z \mid y, z\right\} = E\left\{u \mid y\right\} - \alpha y - \beta z = 0$$

and

$$\alpha < y^2 > + \beta < yz > = < uy > , \quad \alpha < yz > + \beta < z^2 > = < uz >$$

But $\beta = 0$ from the first equation. Substitution into the second and third equation gives (9.6).

If, in addition, the process is stationary, Eq. (9.6) goes over into

$$R\left(t\right) R\left(\tau\right) = R\left(t + \tau\right) R\left(0\right)$$

whence

$$R\left(\tau\right) = R\left(0\right) e^{-\alpha|\tau|} \qquad (9.7)$$

Therefore, the autocorrelation of a stationary, normal ly distributed Markov process must have the exponential form (9.7)! Stationary Gaussian processes will, therefore, in general not be Markovian.

Finding the p.-densities of an arbitrary random process is, in general, a difficult, if not impossible task. For the transition probability of a Markov process, however, a differential equation exists known as Fokker - Planck equation, or Kolmogorov equation.

The starting point for the derivation of this equation is the Chapman - Kolmogorov integral equation (9.3). We generalize it immediately to an n-dimensional Markov process $\mathbf{x}(t) = \left[x_1(t) \ldots x_n(t) \right]$:

$$p(\mathbf{x};t+\tau|\mathbf{x}_0;0) = \int_{-\infty}^{+\infty} \ldots \int_{-\infty}^{+\infty} p(\mathbf{x};t+\tau|\mathbf{y};t)\, p(\mathbf{y};t|\mathbf{x}_0;0)\, dy_1 \ldots dy_n$$

$$(9.8)$$

Multiplication of both sides of this equation with an arbitrary scalar function $R(\mathbf{x})$ which vanishes for $x_1 \to \infty, \ldots x_n \to \infty$, and integration leads to

$$\int_{-\infty}^{+\infty} \ldots \int_{-\infty}^{+\infty} R(\mathbf{x})\, p(\mathbf{x},t+\tau|\mathbf{x}_0;0)\, dx_1 \ldots dx_n =$$

$$= \int_{-\infty}^{+\infty} \ldots \int_{-\infty}^{+\infty} dy_1 \ldots dy_n \int_{-\infty}^{+\infty} \ldots \int_{-\infty}^{+\infty} R(\mathbf{x})\, p(\mathbf{x};t+\tau|\mathbf{y};t)\, p(\mathbf{y},t|\mathbf{x}_0;0)\, dx_1 \ldots dx_n$$

We now expand $R(\mathbf{x})$ in a Taylor series in terms of $(x_i - y_i)$

$$R(\mathbf{x}) = R(\mathbf{y}) + \sum_{i=1}^{n}(x_i-y_i)\frac{\partial R}{\partial y_i} + \frac{1}{2}\sum_{i=1}^{n}\sum_{j=1}^{n}(x_i-y_i)(x_j-y_j)\frac{\partial^2 R}{\partial y_i \partial y_j} + \ldots$$

and substitute into the right-hand side of the preced
ing equation:

$$\int\ldots\int dy_1\ldots dy_n \left[R(\mathbf{y})p(\mathbf{y},t|\mathbf{x}_0;0)\underbrace{\int\ldots\int p(\mathbf{x};t+\tau|\mathbf{y};t)\,dx_1\ldots dx_n}_{1} + \right.$$

$$+\sum_i \frac{\partial R}{\partial y_i}\, p(\mathbf{y};t|\mathbf{x}_0;0)\underbrace{\int\ldots\int(x_i-y_i)\,p(\mathbf{x};t+\tau|\mathbf{y};t)\,dx_1\ldots dx_n}_{a_i(\mathbf{y},t,\tau)} +$$

$$+\frac{1}{2}\sum_i\sum_j \frac{\partial^2 R}{\partial y_i \partial y_j}\, p(\mathbf{y};t|\mathbf{x}_0;0)\underbrace{\int\ldots\int(x_i-y_i)(x_j-y_j)p(\mathbf{x};t+\tau|\mathbf{y};t)\,dx_1\ldots dx_n + \ldots}_{b_{ij}(\mathbf{y},t,\tau)}$$

$$= \int\ldots\int \left[R(\mathbf{x})p(\mathbf{x};t|\mathbf{x}_0;0) + \sum_i a_i(\mathbf{x},t,\tau)\frac{\partial R}{\partial x_i} p(\mathbf{x};t|\mathbf{x}_0;0) + \right.$$

$$\left. + \frac{1}{2}\sum_i\sum_j b_{ij}(\mathbf{x},t,\tau)\frac{\partial^2 R}{\partial x_i \partial x_j}\, p(\mathbf{x};t|\mathbf{x}_0;0) + \ldots\right]dx_1\ldots dx_n$$

where, at the end, we have written $\mathbf{x}$ instead of y .

Integration by parts now renders, for the complete equation,

$$\int \ldots \int R(\mathbf{x}) \left\{ p(\mathbf{x};t+\tau | \mathbf{x}_0;0) - p(\mathbf{x};t | \mathbf{x}_0;0) + \sum_i \frac{\partial}{\partial x_i} \left[a_i\, p(\mathbf{x};t | \mathbf{x}_0;0) \right] - \right.$$

$$\left. - \frac{1}{2} \sum_i \sum_j \frac{\partial^2}{\partial x_i \partial x_j} \left[b_{ij}\, p(\mathbf{x};t | \mathbf{x}_0;0) \right] + \ldots \right\} dx_1 \ldots dx_n = 0$$

Since $R(\mathbf{x})$ is arbitrary the integrand must vanish. If, in addition, we divide by τ and go to the limit $\tau \rightarrow 0$ we obtain the Fokker - Planck equation for the transition probability $p(\mathbf{x};t | \mathbf{x}_0;0)$

$$\frac{\partial p}{\partial t} + \sum_{i=1}^{n} \frac{\partial}{\partial x_i}(\alpha_i p) - \frac{1}{2} \sum_{i=1}^{n} \sum_{j=1}^{n} \frac{\partial^2}{\partial x_i \partial x_j}(\beta_{ij} p) = 0 \qquad (9.9)$$

with the corresponding initial condition

$$p(\mathbf{x};0 | \mathbf{x}_0;0) = \delta(\mathbf{x}-\mathbf{x}_0) \qquad (9.10)$$

Coefficients α_i and β_{ij} are defined by

$$\left. \begin{array}{l} \alpha_i(\mathbf{x},t) = \lim_{\tau \to 0} \frac{1}{\tau} \int_{-\infty}^{+\infty} \ldots \int_{-\infty}^{+\infty} (y_i - x_i) p(\mathbf{y};t+\tau | \mathbf{x};t)\, dy_1 \ldots dy_n \\[4mm] \beta_{ij}(\mathbf{x},t) = \lim_{\tau \to 0} \frac{1}{\tau} \int_{-\infty}^{+\infty} \ldots \int_{-\infty}^{+\infty} (y_i - x_i)(y_j - x_j) p(\mathbf{y};t+\tau | \mathbf{x};t)\, dy_1 \ldots dy_n \end{array} \right\} (9.11)$$

assuming   that all limits exist and that all higher
terms in the Taylor series vanish for $\tau \to 0$, i.e., for
$\mathbf{y} \to \mathbf{x}$, of order $O(\tau^2)$.

Coefficients (9.11) may be written as condi
tional expectations. Upon putting

$$y_i - x_i = \Delta x_i = \int_t^{t+\tau} \dot{x}_i(\lambda)\, d\lambda \qquad (9.12)$$

we have

$$\left.\begin{aligned}
\alpha_i(\mathbf{x},t) &= \lim_{\tau \to 0} E\left\{ \frac{\Delta x_i}{\tau} \,\Big|\, \mathbf{x}(t) \right\} \\[2em]
\beta_{ij}(\mathbf{x},t) &= \lim_{\tau \to 0} E\left\{ \frac{\Delta x_i\, \Delta x_j}{\tau} \,\Big|\, \mathbf{x}(t) \right\}
\end{aligned}\right\} \qquad (9.13)$$

Let the random process $\mathbf{x}(t)$ now be given by
the following system of n first-order differen ial
equations:

$$\dot{x}_i = f_i(x_1, \ldots x_n, t) + \sum_{j=1}^{m} h_{ij}(x_1, \ldots x_n, t)\, w_j(t) \qquad (9.14)$$

where f_i and h_{ij} are given functions of the random
variable $\mathbf{x}$ and of time t, and $w_i(t)$ represents white

noise with zero mean and correlation functions

$$R_{ij}(t,s) = E\left\{w_i(t)\,w_j(s)\right\} = 2\,D_{ij}\,\delta(t-s) \qquad (9.15)$$

Using (9.14) one has from (9.12) for $\tau \rightarrow 0$

$$\Delta x_i = \tau f_i(x_1 \ldots x_n, t) + \sum_{j=1}^{m} \int_{t}^{t+\tau} h_{ij}(x_1 \ldots x_n, \lambda) w_j(\lambda)\,d\lambda$$

Therefore, and because of

$$E\left\{f(x,t)\,\middle|\,x(t)\right\} = f(x,t)$$

it follows from Eq. (9.13)

$$\alpha_i = \lim_{\tau \to 0}\left[f_i(x,t) + \sum_{j=1}^{m} h_{ij}(x,t)\,\frac{1}{\tau}\int_{t}^{t+\tau} E\left\{w_j(\lambda)\right\}\,d\lambda\right] = f_i(x,t)$$

$$\beta_{ij} = \lim_{\tau \to 0}\left[\tau f_i\, f_j + f_i \sum_{k} h_{ik}\int_{t}^{t+\tau} E\left\{w_k(\lambda)\right\}\,d\lambda \right. +$$

$$\left. + f_j \sum_{r} h_{jr}\int_{t}^{t+\tau} E\left\{w_r(\lambda)\right\}\,d\lambda + \sum_{k}\sum_{r} h_{ik}\,h_{jr}\,\frac{1}{\tau}\int_{t}^{t+\tau}d\mu\int_{t}^{t+\tau} E\left\{w_k(\lambda)w_r(\mu)\right\}\,d\lambda\right] =$$

$$= 2\sum_{r}\sum_{k} h_{ik}\,h_{jr}\,D_{kr}\,\underbrace{\lim_{\tau \to 0}\frac{1}{\tau}\int_{t}^{t+\tau}d\mu\int_{t}^{t+\tau}\delta(\lambda-\mu)\,d\lambda}_{1} = 2\sum_{r}\sum_{k} D_{kr}\,h_{ik}\,h_{jr}$$

The Fokker - Planck equation (9.9) then goes over into

$$\frac{\partial p}{\partial t} = -\sum_i \frac{\partial}{\partial x_i}(f_i\, p) + \sum_k \sum_r D_{kr} \sum_i \sum_j \frac{\partial^2}{\partial x_i\, \partial x_j}(h_{ik}\, h_{jr}\, p)$$

$$(9.16)$$

with initial condition

$$p(x_1, \ldots x_n; 0 \mid x_1^0, \ldots x_n^0; 0) = \delta(x_1 - x_1^0) \ldots \delta(x_n - x_n^0) \quad (9.17)$$

For the frequently occurring special case where the system (9.14) is of the form

$$\left. \begin{aligned} \dot{x}_1 &= f_1 + h_{11}\, w \\[2mm] \dot{x}_2 &= f_2 + h_{21}\, w \\[2mm] \dot{x}_3 &= f_3 + h_{31}\, w \end{aligned} \right\} \qquad (9.14a)$$

with

$$R = 2\, D\, \delta(t - s) \qquad (9.15a)$$

instead of (9.15), the Fokker - Planck equation (9.16) reads

$$\frac{\partial p}{\partial t} = -\frac{\partial}{\partial x_1}(f_1\, p) - \frac{\partial}{\partial x_2}(f_2\, p) - \frac{\partial}{\partial x_3}(f_3\, p) +$$

$$(9.16a)$$

$$+ D \sum_{i=1}^{3} \frac{\partial}{\partial x_i} \left[\frac{\partial}{\partial x_1}(h_{i1}\, h_{11}\, p) + \frac{\partial}{\partial x_2}(h_{i1}\, h_{21}\, p) + \frac{\partial}{\partial x_3}(h_{i1}\, h_{31}\, p) \right]$$

The derivation of the Fokker-Planck equa
tion given above is purely formal, as is Eq. (9.14).
Indeed, since the right-hand side of this equation
contains white noise $\dot{x}_i$ does not exist: the process
is not differentiable. Nevertheless, the equations
can be used formally as may be seen employing the
Wiener process as an example. Writing Eq. (7.5) formally as

$$\dot{x} = w(t) \qquad\qquad (9.18)$$

one obtains from (9.16) the correct Eq. (7.6) derived
by an entirely different procedure.

The exact derivation of Eq. (9.16) would
involve so-called Ito integrals (also known as stoch
astic integrals) whose properties are basically different from those of ordinary integrals which, if
they contain the differential $dw(t)$ of white noise,
simply do not exist. However, as has been pointed
out already on p 54, random processes appearing in
practical applications are always differentiable.
Pathological functions like white noise are physical
ly not realizable. Only processes $z_i(t)$ with a relative
ly wide frequency band are possible. Therefore, as
has been pointed out by Caughey and Gray [12], it

seems more reasonable to employ ordinary integrals
rather than Ito integrals, and to replace Eq. (9.16)
by the following slightly different equation called
"physical" Fokker-Planck equation

$$\frac{\partial p}{\partial t} = -\sum_i \frac{\partial}{\partial x_i}(f_i p) + \sum_k \sum_r D_{kr} \sum_i \sum_j \frac{\partial}{\partial x_i}\left[h_{ik} \frac{\partial}{\partial x_j}(h_{jr} p)\right] \qquad (9.19)$$

The coefficients of intensity D_{ij} are to be determined from

$$2 D_{ij} = \int_{-\infty}^{+\infty} E\left\{z_i(t) z_j(t+\tau)\right\} d\tau \qquad (9.20)$$

Finally, it should be remarked that a
Fokker-Planck equation may also be set up for discrete processes $x(t)$ which can take on only the value
a_0, a_1 ... a_n ...,[4] ,p. 543. We give here the resu
Employing, as in Eqs. (9.13), the notation

$$P\left[x(t) = a_i \mid x(\tau) = a_j\right] = p_{ij}(t,\tau) \qquad (9.21$$

then

$$\sum_i p_{ij}(t,\tau) = 1, \quad p_{ij}(\tau,\tau) = \delta_{ij} \qquad (9.22)$$

and one obtains the Fokker-Planck equation

$$\frac{\partial p_{ij}}{\partial t} = - q_i(t)\, p_{ij} + \sum_{k \neq i} q_{ik}(t)\, p_{kj} \tag{9.23}$$

where

$$\left. \begin{array}{l} P\Big[x(t+s)=a_i \,\big|\, x(t)=a_j\Big] = q_{ij}(t)\, s \quad (i \neq j) \\[12pt] P\Big[x(t+s)=a_i \,\big|\, x(t)=a_i\Big] = 1 - q_i(t)\, s \end{array} \right\} \begin{array}{l} \text{for } s \to 0 \\[12pt] (9.24) \end{array}$$

Hence

$$q_i(t) = \sum_{j \neq i} q_{ji}(t) \tag{9.25}$$

As an example, consider the Poisson process. Here $a_n = n$, and one notes that $x(t)$ can only increase, at most by 1 in the time interval $s \to 0$. Therefore, comparing Eqs. (6.1) with (9.24) it follows that

$$q_i(t) = q_{i,i-1}(t) = \nu,$$

with all other zero. The Fokker-Planck equation (9.23) reads therefore, with $p_{i0}(t,0) \equiv p_i(t)$,

$$\frac{dp_i}{dt} = - \nu p_i + \nu p_{i-1}$$

in accordance with Eq. (6.5).

Literature .

[1] A.N. Kolmogorov: Ueber die analytischen Methoden in der Wahrscheinlichkeitsrechnung. Math. Ann.104 (1931),p. 415.

[2] A.N. Kolmogorov: Grundbegriffe der Wahrscheinlichkeitsrechnung. Ergeb. Math. und ihrer Grenzgeb. Bd.2, No.3, 1933.

[3] R. von Mises: Wahrscheinlichkeitsrechnung. Deuticke. Wien 1931.

[4] A. Papoulis: Probability, Random Variables and Stochastic Processes. McGraw-Hill Book Comp. New York 1965.

[5] E. Parzen: Modern Probability Theory and its Applications. Wiley. New York 1960.

[6] E. Parzen: Stochastic Processes. Holden-Day. San Francisco 1962.

[7] J. Heinhold und K.-W. Gaede: Ingenieur-Statistik. Oldenbourg-Verlag. München-Wien 1964.

[8] A.A. Sweschnikow: Untersuchungsmethoden der Theorie der Zufallsfunktionen. Teubner. Leipzig 1965.

[9]   W.Eberl: Einführung in die Stochastik. I. Teil,
      Wahrscheinlichkeitstheorie. Wiener Schwachstrom-
      werke, Wien 1965.

[10] M.F.M. Osborne: Reply to comments on Brownian
      motion in the stock market. Oper. Research 7
      (1959), 807-811.

[11] E. Wax (editor): Selected Papers on Noise and
      Stochastic Processes. Dover Publications. New
      York 1954.

[12] A.H. Gray and T.K. Caughey: A controversy in prob
      lems involving random parametric excitation.
      J. Math. Phys. 44 (1965),288.

C H A P T E R   I I

LINEAR RANDOM VIBRATIONS .

The following treatement will be restrictive in the sense of correlation theory.

1. The Oscillator with One Degree of Freedom under Stochastic Excitation .

This is the simplest case. The corresponding differential equation reads

$$\ddot{y} + 2\,\xi\,\omega_0\,\dot{y} + \omega_0^2 y = x(t)\,H(t) \tag{1.1}$$

where $x(t)$ is a stationary random driving function
with mean $m_x(t) = 0$.

For the admittance one has at once

$$F(\omega) = \frac{1}{\omega_0^2 - \omega^2 + 2\,i\,\xi\,\omega_0\,\omega} \tag{1.2}$$

and for the impulse response

$$h(t) = \frac{1}{\omega_0\,\sqrt{1-\xi^2}}\;e^{-\xi\omega_0 t}\;\sin\left(\omega_0\,\sqrt{1-\xi^2}\,t\right) \tag{1.3}$$

Eq. (I-5.30) renders $m_y = 0$ for the expectation of
$y(t)$. The spectral density of the stationary solu

tion is, from Eq. (I - 5.38),

$$S_{yy}(\omega) = \frac{S_{xx}(\omega)}{(\omega_0^2 - \omega^2)^2 + 4\,\xi^2\omega_0^2\,\omega^2} \qquad (1.4)$$

and the corresponding autocorrelation is, according to (I - 5.17),

$$R_{yy}(\tau) = \frac{1}{\pi} \int_0^{+\infty} \frac{S_{xx}(\omega)}{(\omega_0^2 - \omega^2)^2 + 4\,\xi^2\omega_0^2\,\omega^2} \cos \omega\tau \, d\omega \qquad (1.5)$$

We consider the special case of white noise excitation, $x(t) = A\,w(t)$. From the table on p. 52 we have $S_{xx}(\omega) = A^2$. If this is substituted into (1.5) we obtain, after integration,

$$R_{yy}(\tau) = A^2 \frac{e^{-\xi\omega_0\tau}}{4\,\xi\,\omega_0^3} \left[\cos\left(\omega_0\sqrt{1-\xi^2}\,\tau\right) + \frac{\xi}{\sqrt{1-\xi^2}} \sin \omega_0 \sqrt{1-\xi^2}\,\tau \right] \qquad (1.6)$$

whence, for $\tau = 0$

$$\sigma_y^2 = \frac{A^2}{4\,\xi\,\omega_0^3} \qquad (1.7)$$

During the transient state, $t < \infty$, $y(t)$ is a nonstationary random function. Its autocorrelation follows from Eq. (I - 5.32) upon substitution of $R_{xx}(t) = A^2\,\delta(t)$.

Assuming $t \geq s$ we have

$$R_{yy}(t,s) = A^2 \int_0^t \int_0^s h(t-\tau) \, h(s-\lambda) \, \delta(\lambda-\tau) \, d\lambda \, d\tau =$$

$$= A^2 \int_0^t h(t-\tau) \, h(s-\tau) \, d\tau$$

$$R_{yy}(t,s) = \frac{A^2 e^{-\xi\omega_0 t}}{4\omega_0} \left\{ \frac{e^{+\xi\omega_0 s}}{\xi\omega_0^2} \left[\cos\left(\omega_0\sqrt{1-\xi^2}(t-s)\right) + \frac{\xi}{\sqrt{1-\xi^2}} \sin\left(\omega_0\sqrt{1-\xi^2}\,(t-s)\right) \right] + \right.$$

$$+ \frac{e^{-\xi\omega_0 s}}{\omega_0^2(1-\xi^2)} \left[\xi \cos\left(\omega_0\sqrt{1-\xi^2}(t+s)\right) - \sqrt{1-\xi^2} \sin\left(\omega_0\sqrt{1-\xi^2}\,(t+s)\right) - \right.$$

$$\left. \left. - \frac{1}{\xi} \cos\left(\omega_0\sqrt{1-\xi^2}\,(t-s)\right) \right] \right\}$$

(1.8)

This renders for the variance

$$\sigma_y^2(t) = \frac{A^2}{4\xi\omega_0^3} + \frac{A^2 e^{-2\xi\omega_0 t}}{4\omega_0^3(1-\xi^2)} \left[\xi \cos\left(2\omega_0\sqrt{1-\xi^2}\,t\right) - \sqrt{1-\xi^2} \sin\left(2\omega_0\sqrt{1-\xi^2}\,t\right) - \frac{1}{\xi} \right]$$

(1.9)

which, for $t \to \infty$, goes over into (1.7).

White noise represents the extreme case of a pure random function. It fluctuates with infinitely large amplitude infinitely often in each finite time-interval, thus producing an extremely violent excita‌tion of the oscillator. It is, therefore, not surpris‌

ing that oscillations can be kept within limits by strong damping only. We note from (1.7) and (1.9), putting $\zeta = 0$, that the variance of the undamped oscil_ lator grows beyond all limits as $t \longrightarrow \infty$.

If the input $x(t)$ is normally distributed the output $y(t)$ is normal. In that case the p.-density of $y(t)$ is completely determined by $m_y(t)$ and $R_{yy}(t,s)$, cf. Eq. (I-2.35).

2. System with n Degrees of Freedom.

We consider a holonomic system of n degrees of freedom, with corresponding generalized co ordinates $q_i(t)$. Let $q_i = 0$ in the equilibrium posi tion.

It is convenient, by introducing the co ordinates

$$\left. \begin{array}{l} y_i = q_i \quad (i = 1, 2, \ldots n) \\ y_i = \dot{q}_{i-n} \quad (i = n+1, \ldots 2n) \end{array} \right\} \qquad (2.1)$$

to transform the Lagrangian equations of motion into the form

$$\dot{y}(t) = Py(t) + f(t) \qquad (2.2)$$

where $y = (y_1, y_2 \ldots \ldots y_{2n})$, P is a $2n \times 2n$ coefficient matrix, and $f(t)$ represents the vector $(0,0 \ldots 0, Q_1(t), Q_2(t) \ldots Q_n(t))$ of the driving forces. We suppose that P is nonsingular and that its $2n$ eigenvalues are different, with negative real part. Equilibrium $y_i = 0$ is, therefore, stable.

Let the principal-axes-transformation of (2.2) be given by the nonsingular $2n \times 2n$ matrix T

$$T^{-1}PT = D = Diag(\varrho_1, \ldots \varrho_{2n}) \qquad (2.3)$$

where $\varrho_1 \ldots \varrho_{2n}$ are the eigenvalues of P. Putting

$$y(t) = Tz(t) \quad or \quad y_i(t) = T_{ij} z_j(t) \qquad (2.4)$$

Eq. (2.2), after multiplication from the left by T^{-1}, goes over into

$$\dot{z}(t) = Dz(t) + x(t) \qquad (2.5)$$

where $x(t) = T^{-1}f(t)$, or $x_i(t) = T_{ij}^{-1} f_j(t)$

The solution of (2.5) corresponding to $z_i(0) = 0$ is given by

$$\left. \begin{aligned} z(t) &= \int_0^t H(t-\tau)x(\tau)d\tau \\ \\ with \quad H(t) &= e^{Dt} = Diag(e^{\varrho_1 t}, \ldots e^{\varrho_{2n} t}) \end{aligned} \right\} \qquad (2.6)$$

We have, therefore,

$$z_i(t) = \int_0^t e^{\varrho_i(t-\tau)} x_i(\tau) d\tau \qquad (2.7)$$

Analysis now continues as in the case of one degree of freedom. It is only necessary to determine the correlation functions of the $x_i(t)$ from those of the $f_i(t)$ or $Q_{i-n}(t)$. One easily finds the following equation for correlation and spectral density of a sum $u(t) = x(t) + y(t)$[*]:

$$R_{uu}(\tau) = R_{xx}(\tau) + R_{yy}(\tau) + R_{xy}(\tau) + R_{yx}(\tau) \qquad (2.8)$$

The same is true for the spectral densities. If the processes x and y are orthogonal, $R_{xy} = R_{yx} = 0$ and $S_{xy} = S_{yx} = 0$.

---

[*] For a sum of n variables $u(t) = \sum_{i=1}^{} x_i(t)$ the analogous relation holds

$$R_{uu}(\tau) = \sum_{i=1}^{n} \sum_{j=1}^{n} R_{ij}(\tau) \qquad (2.9)$$

3. Continuous Systems .

Bending   of   a   Beam.

With the aid of the Ritz method a continuous system may always be reduced approximately to one
with a finite number of degrees of freedom. Choosing
for the corresponding functions in the Ritz series
the eigenfunctions , kinematic and dynamic boundary
conditions are automatically satisfied and one can
be sure to get arbitrarily close to the exact solution by increasing the number of terms.

As an example we consider the beam on two
supports, Fig. 2.1,
under the action of a
load $q(x,t)$, distrib
uted randomly with re-
spect to space and time.
The material of the beam
is assumed to possess
internal damping follow
ing the Voigt-Kelvin

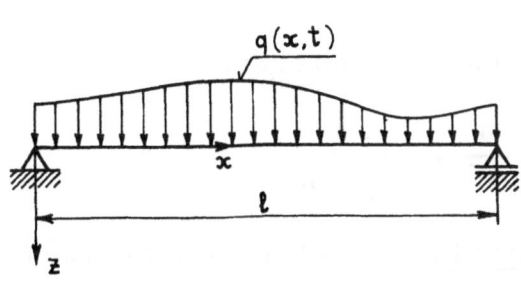

Fig.2.1

constitutive equation of viscoelasticity. After inte
gration over the cross-section of the beam this law
may be written as a linear relation between bending
moment M and rotation φ of the cross-section [1]

$$M = - E J \left(1 + \varepsilon \; \frac{\partial}{\partial t} \right) \frac{\partial \varphi}{\partial x} \qquad (3.1)$$

where ε is the damping coefficient, of dimension of time t. If the Bernoulli-Euler model is used for the beam, i.e., if the influence of shearing forces and rotatory inertia is neglected, then $\varphi = \partial w / \partial x$, where $w(x,t)$ represents deflection, and one obtains, substituting M from the law of angular momentum

$$\frac{\partial^2 M}{\partial x^2} = \varrho F \; \frac{\partial^2 w}{\partial t^2} - q$$

the following equation of the viscous beam

$$c^2 \left(1 + \varepsilon \; \frac{\partial}{\partial t} \right) \frac{\partial^4 w}{\partial x^4} + \frac{\partial^2 w}{\partial t^2} = \frac{q}{\varrho F} \; , \quad c^2 = \frac{E J}{\varrho F} \qquad (3.2)$$

The eigenfunctions will be determined first. Putting, for the n-th eigenfunction,

$$w_n^e = A_n \, e^{\beta_n t} \sin \lambda_n x \quad , \quad \lambda_n = \frac{n \pi}{\ell} \qquad (3.3)$$

boundary conditions are satisfied, and one obtains from (3.2), with $q \equiv 0$, the following frequency equation

$$\beta_n^2 + \varepsilon c^2 \lambda_n^4 \beta_n + c^2 \lambda_n^4 = 0 \qquad (3.4)$$

Separating into real and imaginary parts

$$\beta_n = \alpha_n + i\omega_n \quad , \tag{3.5}$$

one gets

$$\alpha_n = -\frac{\varepsilon c^2}{2}\lambda_n^4 \, , \quad \omega_n = c^2\lambda_n^4\left(1 - \frac{\varepsilon^2 c^2}{4}\lambda_n^4\right) \tag{3.6}$$

Admittances for $w(x,t)$ and $M(x,t)$ may now be determined. Putting

$$q(x,t) = e^{i\omega t}\sin\lambda_n x \, , \quad w_n^P(x,t) = F_n(\omega)e^{i\omega t}\sin\lambda_n x$$

$$\left. M_n^P(x,t) = G_n(\omega)e^{i\omega t}\sin\lambda_n x \right\} \tag{3.7}$$

we find, upon substitution into (3.2),

$$F_n(\omega) = \frac{1}{\varrho F D(\omega)} \, , \quad G_n(\omega) = -\frac{1 + i\omega\varepsilon}{D(\omega)} \, c^2\lambda_n^2$$

$$\left. D(\omega) = -\omega^2 + i\varepsilon c^2\lambda_n^4\omega + c^2\lambda_n^4 \right\} \tag{3.8}$$

Now we concentrate on the study of the random vibrations. We first expand $q(x,t)$ in terms of the eigenfunctions [*]

$$q(x,t) = \sum_{n=1}^{\infty} q_n(t)\sin\lambda_n x \tag{3.9}$$

---

[*] Compare also [2].

where

$$q_n(t) = \frac{2}{\ell} \int_0^\ell q(x,t) \sin \lambda_n x \, dx \qquad (3.10)$$

For each component $w_n(x,t)$, Eq. (I-5.20) holds[*].
Summing

$$w(x,t) = \sum_{n=1}^\infty w_n(x,t) = \sum_{n=1}^\infty \sin \lambda_n x \int_{-\infty}^{+\infty} h_n(t-\tau) q_n(\tau) d\tau \qquad (3.11)$$

Substitution of (3.10) renders

$$w(x,t) = \frac{2}{\ell} \sum_{n=1}^\infty \sin \lambda_n x \int_{-\infty}^{+\infty} h_n(t-\tau) \int_0^\ell q(\xi,\tau) \sin \lambda_n \xi \, d\xi \, d\tau \qquad (3.12)$$

For simplicity, we assume that $E\{q(x,t)\} = 0$.
Then $E\{w(x,t)\} = 0$, and we get for the correlation
function of w, cf. (I-5.32),

$$E\{w(x_1,t) \cdot w(x_2,s)\} = \frac{4}{\ell^2} \sum_{n=1}^\infty \sum_{m=1}^\infty \sin \lambda_n x_1 \sin \lambda_m x_2 \int_{-\infty}^{+\infty} \int_{-\infty}^{+\infty} h_n(t-\tau) h_m(s-\mu) \cdot$$

$$\left[\int_0^\ell \int_0^\ell E\{q(\xi,\tau) \, q(\eta,\mu)\} \sin \lambda_n \xi \sin \lambda_m \eta \, d\xi \, d\eta \right] d\tau \, d\mu \qquad (3.13)$$

---

[*] We restrict our attention to stationary motion!

This completes the solution of the problem, at least in principle, in the sense of correlation theory. The actual evaluation of (3.13) for a given correlation function $E\{q(\xi,\tau) \cdot q(\eta,\mu)\}$ will, in general, present considerable numerical difficulties.

The problem is simplified drastically if one assumes that the load is not correlated spacewise,

$$E\{q(\xi,\tau)q(\eta,\mu)\} = \frac{\ell}{2} R(\tau-\mu)\delta(\xi-\eta) \qquad (3.14)$$

The second of the two double - integrals may then be evaluated in closed form. As a consequence of the orthogonality of the circular functions it vanishes for $m \neq n$, and the double sum reduces to a simple sum. Putting, in addition, $\tau = t - u$, $\mu = s - v$, we get

$$E\{w(x_1,t+\tau)w(x_2,t)\} = \sum_{n=1}^{\infty} \sin \lambda_n x_1 \sin \lambda_n x_2 \int_{-\infty}^{+\infty} dv \left[\int_{-\infty}^{+\infty} h_n(u) h_n(v) R(\tau+v-u)du \right]$$

$$= \sum_{n=1}^{\infty} \sin \lambda_n x_1 \sin \lambda_n x_2 R_n(\tau) = \frac{1}{\pi} \sum_{n=1}^{\infty} \sin \lambda_n x_1 \sin \lambda_n x_2 \int_0^{\infty} S_n(\omega)\cos \omega\tau \, d\omega$$

$$(3.15)$$

As was to be expected, the correlation function of w depends on τ but not on t. According to

Eq. (I-5.32), the function $R_n(\tau)$ is the time-part of
the correlation function of $w_n(x,t)$. The latter is
connected with the corresponding spectral density
$S_n(\omega)$ through Eq. (I-5.17).

A similar expression may be given for
the correlation function of the bending moment $M(x,t)$.

The spectral density $S_n(\omega)$ is connect‐
ed with the spectral density of the excitation via
Eq. (I-5.38). One has, therefore, finally,

$$E\{w(x_1,t+\tau).w(x_2,t)\} = \sum_{n=1}^{\infty} \sin \lambda_n x_1 \sin \lambda_n x_2 \int_0^{\infty} S(\omega)|F_n(\omega)|^2 \cos \omega \tau \, d\omega$$

(3.16)

where $S(\omega)$ is the excitation spectral density, cor‐
responding to $R(t)$.

An identical expression is obtained
for the bending moment, with $G_n(\omega)$ replacing $F_n(\omega)$.

The integral in (3.16) can only be eval‐
uated (in general numerically) if $S(\omega)$ is given. The
evaluation is possible in closed form in the extreme
case of white noise excitation. It turns out, however
that the variance of the bending moment then becomes
infinite, while the variance of the deflection re‐
mains finite. If, on the other hand, shearing stiff‐
ness and rotatory inertia are taken into account
(Timoshenko beam), both variance of deflection and
variance of bending moment remain finite [1].

Literature .

[1] S.H. Crandall and A. Yildiz: Random vibration
of beams . J. Appl. Mech. 29 (1962), 267.

[2] W.E. Boyce: Stochastic nonhomogeneous Sturm-
-Liouville problems. J. Franklin Inst. 282
(1966), 206.

C H A P T E R   III

NON-LINEAR RANDOM VIBRATIONS .

Exact solutions are, in general, not available. For finding approximate solutions the same methods as in the deterministic case are being used. We discuss some examples.

1. Oscillator with Nonlinear Restoring Force .

Let the corresponding differential equation be given as

$$\ddot{y} + \alpha \dot{y} + F(y) = x(t) H(t) \qquad (1.1)$$

Assuming $x(t)$ to represent white noise with zero mean we have $y(t)$ a nonstationary Markov process with Fokker-Planck equation (I-9.16) which, in the present case, is identical with (I-9.19),

$$\frac{\partial p}{\partial t} = \frac{\partial}{\partial \dot{y}} \left[(F(y) + \alpha \dot{y}) p \right] - \frac{\partial}{\partial y} (\dot{y} p) + D \frac{\partial^2 p}{\partial \dot{y}^2} \qquad (1.2)$$

for the transition probability $p(y, \dot{y}; t \mid y_0, \dot{y}_0; 0)$. The autocorrelation of $x(t)$ is

$$R_x(\tau) = 2 D \delta(\tau) \qquad (1.3)$$

The general solution of equation (1.2) is not known, at least not in closed form. However, $y(t)$ becomes stationary for $t \to \infty$. Putting, $\partial p / \partial t = 0$ the solution of (1.2) is then [1]

$$p(y,\dot{y}) = C \exp \left\{ - \frac{\alpha}{D} \left[\frac{\dot{y}^2}{2} + V(y) \right] \right\} \qquad (1.4)$$

where C is a normalizing constant, and

$$V(y) = \int_0^y F(\eta) \, d\eta \qquad (1.5)$$

represents the potential of the restoring force F.

The expression within the brackets in Eq. (1.4) is the sum of kinetic and potential energy per unit mass. We note that $p(y,\dot{y})$ is Gaussian in $\dot{y}$. Also, since

$$p(y,\dot{y}) = p(y) \, p(\dot{y})$$

$y(t)$ and $\dot{y}(t)$ are independent.

For the stationary expectation of $y(t)$ one has

$$E\{y(t)\} = C \int_{-\infty}^{+\infty} e^{-\frac{\alpha}{2D} \dot{y}^2} d\dot{y} \int_{-\infty}^{+\infty} y \exp \left\{ - \frac{\alpha}{D} V(y) \right\} dy$$

and, upon integration,

$$E\{y(t)\} = C\sqrt{2\pi\frac{D}{\alpha}} \int_{-\infty}^{+\infty} y\,\exp\left\{-\frac{\alpha}{D}V(y)\right\}dy = m_y \qquad (1.6)$$

If $V(y)$ is an even function $E\{y(t)\}$ vanishes.

Similarly, for the variance

$$\sigma_y^2 = C\sqrt{2\pi\frac{D}{\alpha}} \int_{-\infty}^{+\infty} y^2\,\exp\left\{-\frac{\alpha}{D}V(y)\right\}dy - m_y^2 \qquad (1.7)$$

The preceding analysis may be immediately generalized to a system with n degrees of freedom:

$$\ddot{y}_i + \alpha_i\dot{y}_i + \frac{1}{m_i}\frac{\partial V}{\partial y_i} = x_i(t) \qquad (i = 1,2\ldots n) \quad (1.8)$$

The stationary Fokker-Planck equation now reads

$$\sum_{i=1}^{n}\frac{\partial}{\partial\dot{y}_i}\left[\left(\frac{1}{m_i}\frac{\partial V}{\partial y_i}+\alpha_i\dot{y}_i\right)p\right] - \sum_{i=1}^{n}\frac{\partial}{\partial y_i}(\dot{y}_i\,p) + \sum_{i=1}^{n}D_i\frac{\partial^2 p}{\partial\dot{y}_i^2} = 0 \quad (1.9)$$

where

$$<x_i(t)> = 0, \quad <x_i(t)\,x_i(t+\tau)> = 2D_i\delta(\tau), <x_i(t)\,x_j(t+\tau)> = 0 \text{ for } i\neq j$$

has been assumed. Eq. (1.9) has the solution

$$p(y_1,\ldots y_n, \dot{y}_1\ldots\dot{y}_n) = C\exp\left\{-\frac{1}{K}\left(\frac{1}{2}\sum m_i\dot{y}_i^2 + V\right)\right\} \quad (1.10)$$

provided

$$\frac{D_i}{\alpha_i} \, m_i = K$$

With the aid of the preceding formulas a great number of the problems discussed in Chapter IV and related to material fatigue may be solved.

2. Perturbation Method .

The method [2] is applied exactly as in the deterministic case. Let the differential equation be

$$\ddot{x} + 2\zeta\omega_0\dot{x} + \omega_0^2 \left[x + \varepsilon g(x) \right] = F(t) \qquad (2.1)$$

and let the nonlinear part $\varepsilon g(x)$ be small[*]. ε is a smallness parameter. We expand the solution in powers of ε

$$x(t) = x_0(t) + \varepsilon x_1(t) + \varepsilon^2 x_2(t) + \ldots \qquad (2.2)$$

substitute, and equate the coefficients of ε^n to zero. We obtain then the following recursive chain of linear differential equations

---

[*] The same assumption has to be made if equivalent linearization is used, sec. 3.

$$\ddot{x}_0 + 2\xi\omega_0\dot{x}_0 + \omega_0^2 x_0 = f(t)$$

$$\ddot{x}_1 + 2\xi\omega_0\dot{x}_1 + \omega_0^2 x_1 = -\omega_0^2 g(x_0)$$

$$\ddot{x}_2 + 2\xi\omega_0\dot{x}_2 + \omega_0^2 x_2 = -\omega_0^2 x_1 g'(x_0)$$

$$- - - - - - - - - - - - - - - - - - \qquad (2.3)$$

differentiability of $g(x)$ being assumed:

$$g(x_0 + \epsilon x_1 + \epsilon^2 x_2 + \dots) = g(x_0) + \epsilon x_1 g'(x_0) + \dots$$

Employing Eq.(I-5.20) we have

$$x_0(t) = \int_0^\infty h(\tau) f(t-\tau) d\tau \qquad (2.4)$$

and

$$x_1(t) = -\omega_0^2 \int_0^\infty h(\tau) g[x_0(t-\tau)] d\tau \qquad (2.5)$$

and so on

where, in the present case, the impulse response of the system (2.3) is given by

$$h(t) = \frac{\sin \omega_0 \sqrt{1-\xi^2} t}{\omega_0 \sqrt{1-\xi^2}} e^{-\xi\omega_0 t} \qquad (2.6)$$

In principle, all moments of arbitrary order may now be determined. Integration difficulties soon become, however, insurmountable.

As an example, we discuss the Duffing

equation, $g(x) = x^3$, and restrict the solution to a first-order perturbation: $x = x_0 + \varepsilon x_1$. The expectation of x^2 is then, to this order,

$$<x^2> = <x_0^2> + 2\varepsilon <x_0 x_1> \qquad (2.7)$$

From (2.4), upon squaring and taking expectations,

$$<x_0^2> = \int\limits_0^\infty \int\limits_0^\infty h(\tau_1) h(\tau_2) <f(t-\tau_1) \, f(t-\tau_2)> d\tau_1 \, d\tau_2 \qquad (2.8)$$

We assume $f(t)$ to be stationary. Then

$$<f(t-\tau_1) f(t-\tau_2)> = R_f(\tau_2 - \tau_1) \qquad (2.9)$$

Furthermore, from (2.5),

$$<x_0 x_1> = -\omega_0^2 \int\limits_0^\infty h(\tau) <x_0(t) \, x_0^3(t-\tau)> d\tau$$

and, using (2.4),

$$<x_0 x_1> = -\omega_0^2 \int\limits_0^\infty h(\tau) \int\limits_0^\infty h(\tau_1) \int\limits_0^\infty h(\tau_2) \int\limits_0^\infty h(\tau_3) \int\limits_0^\infty h(\tau_4) H \, d\tau \, d\tau_1 \, d\tau_2 \, d\tau_3 \, d\tau_4$$

$$(2.10)$$

where

$$H = <f(t-\tau_1) f(t-\tau-\tau_2) f(t-\tau-\tau_3) f(t-\tau-\tau_4)>$$

The fivefold integral simplifies somewhat for $f(t)$ <u>normal</u> with zero expectation and autocorrelation $R_f(t)$.

Then

$$< f(t) > = 0 \ , \ < f(t_1) f(t_2) > = R_f(t_1 - t_2)$$

$$< f(t_1) f(t_2) f(t_3) > = 0$$

$$\left. \begin{array}{c} \\ \\ \\ \\ \\ \\ \end{array} \right\} \quad (2.11)$$

$$< f(t_1) f(t_2) f(t_3) f(t_4) > = R_f(t_1 - t_2) R_f(t_3 - t_4) +$$

$$+ R_f(t_1 - t_3) R_f(t_2 - t_4) + R_f(t_1 - t_4) R_f(t_2 - t_3)$$

and we have

$$H = R_f\left(\tau - \tau_1 + \tau_2\right) R_f\left(\tau_3 - \tau_4\right) + R_f\left(\tau - \tau_1 + \tau_3\right) R_f\left(\tau_2 - \tau_4\right) +$$

$$+ R_f\left(\tau - \tau_1 + \tau_4\right) R_f\left(\tau_2 - \tau_3\right) \quad\quad\quad (2.12)$$

It is expedient to first determine the autocorrelation of x_0

$$R_0(\tau) = < x_0(t) x_0(t+\tau) > = \int_0^\infty \int_0^\infty h(\tau_1) h(\tau_2) R_f(\tau - \tau_1 + \tau_2) d\tau_1 d\tau_2 \quad (2.13)$$

which also renders

$$< x_0^2 > = R_0(0) = \sigma_0^2$$

Substituting now (2.12) into (2.10), interchanging the order of integration and using (2.13), one obtains

$$< x_0 x_1 > = - 3 \omega_0^2 \sigma_0^2 \int_0^\infty h(\tau) R_0(\tau) d\tau \quad\quad (2.14)$$

Eq. (2.7) then gives

$$<x^2> = \sigma_0^2 \left[1 - 6\varepsilon\omega_0^2 \int_0^\infty h(\tau) R_0(\tau) d\tau \right] \qquad (2.15)$$

If $f(t)$ is Gaussian white noise with $R_f(\tau) =$ $= C\delta(\tau)$, one finds, after some manipulation, using Eq. (2.6),

$$<x^2> = \sigma_0^2 \left[1 - 3\varepsilon\sigma_0^2 \right] \qquad (2.16)$$

We note the reduction of the variance for a "hard" spring, $\varepsilon > 0$.

3. Equivalent Linearization .

The method, [3] and [4], will be demonstrated using the oscillator with nonlinear restoring force as an example. The corresponding differential equation is

$$\ddot{x} + 2\zeta\omega_0\dot{x} + \omega_0^2 x + g(x) = f(t) \qquad (3.1)$$

with $f(t)$ a stationary random function with zero mean.

We restrict discussion to the stationary solution and replace (3.1) by the "equivalent" linear equation

$$\ddot{x} + 2\zeta\omega_0\dot{x} + \omega_0^2 x + \lambda x = f(t) \qquad (3.2)$$

In the deterministic case, the function (error)

$$F(x) = \lambda x - g(x)$$

is "balanced" in the lowest harmonic by introducing a periodic solution and expanding $g(x)$ in a Fourier series. With $f(t)$ and, hence, $x(t)$ random functions this is no longer possible. Instead, we require now that the variance of the function $F(x)$ be a minimum

$$E\left\{F^2(x)\right\} = E\left\{\lambda^2 x^2 - 2\lambda x g(x) + g^2(x)\right\} \longrightarrow \text{Min}$$

From

$$\frac{\partial E\left\{F^2\right\}}{\partial \lambda} = 0$$

we then obtain

$$\lambda = \frac{E\left\{x g(x)\right\}}{E\left\{x^2\right\}} \qquad (3.3)$$

Neither numerator nor determinator are known in Eq. (3.3). To determine them in first approximation the linearized equation (3.2) may be used. We have then

$$R_x(0) = E\left\{x^2\right\} = \frac{1}{2\pi} \int_{-\infty}^{+\infty} \frac{S_f(\omega)}{\left|\lambda + \omega_0^2 - \omega^2 + 2i\xi\omega\omega_0\right|^2} d\omega \qquad (3.4)$$

With spectral density of the driving function $f(t)$ given, the integral may be evaluated using the calcu

lus of residues. We still need $E\{xg(x)\}$. If $f(t)$
is normally distributed, then so is x within first
approximation. With $m = 0$ and $E\{x^2\} = \sigma_x^2$ its probability distribution is given by $N(0,\sigma_x^2)$, and we have

$$E\{xg(x)\} = \int_{-\infty}^{+\infty} xg(x)\, p(x)\, dx \qquad (3.5)$$

The more general case, where nonlinearities
appear also in $\dot{x}$

$$\ddot{x} + 2\zeta\omega_0\dot{x} + \omega_0^2 x + g(x,\dot{x},t) = f(t)$$

and where g depends explicitly on t has been treated
by Caughey [4] .

Literature .

[1]   T.K. Caughey: Derivation and application of the
Fokker-Planck equation. J. Acoust.Soc.America
35 (1963), 1683.

[2]   S.H. Crandall: Perturbation techniques for ran
dom vibration of nonlinear systems. J. Acoust.
Soc. America 35 (1963), 1700.

[3]   S.H. Crandall (editor): Random Vibrations. Vol.2
p.97, M.I.T. Press 1963.

[4]   T.K. Caughey: Equivalent linearization techniques. J. Acoust. Soc. America 35 (1963),
1706.

C H A P T E R   IV

ZERO CROSSINGS, PEAKS AND MATERIAL FATIGUE .

To determine the life expectancy of a
structure or the probability of its failure, the probability for the stress to exceed a given value,[*] as
well as of the mean frequency of this exceedance, must
be known. In addition, it will be important to know
the distribution of subsequent stress peaks. All these
questions are, in general, difficult to answer, particularly so for nonlinear structures.

1. Exceedance of a Given Value.

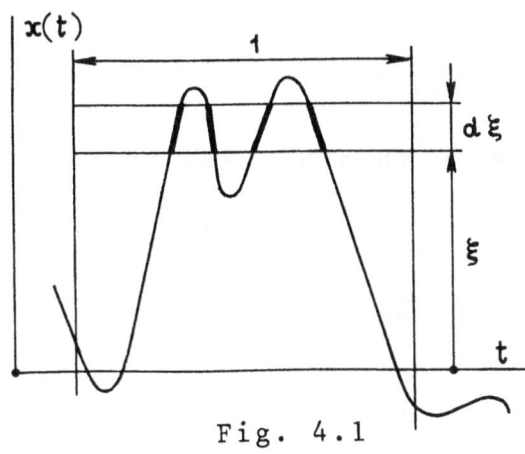

Fig. 4.1

Consider the stationary
random function $x(t)$ and
its time-derivative $\dot{x}(t)$.
Let the joint probabili-
ty density of x and $\dot{x}$
be $p(\xi,\eta)$, [1].

The probability
that x at time t lies
between ξ and $\xi + d\xi$

---

[*] See also Chap. V, section 8. A generalization is
given in [5] .

and at the same time, $\dot{x}$ lies between η and $\eta+d\eta$ or, in other words, that the strip between ξ and $\xi + d\xi$ is being crossed with speed between η and $\eta + d\eta$, cf. Fig. 4.1, is given by $p(\xi,\eta)d\xi\,d\eta$. On the other hand, this expression also represents the time interval which, during unit time, is being spent[*] by x with in the strip $d\xi$ while moving with speed η. Now, the duration of a single crossing of the strip is given by $d\xi/|\eta|$, where the absolute sign has to be taken since it is immaterial whether the speed η is positive (going upwards) or negative (going downwards). Dividing now the total time being spent within the strip by the duration of one crossing we obtain the mean number of crossings per unit time, with speed η, (crossing frequency) of the line $x = \xi$:

$$n_{\xi,\eta} = |\eta|\, p(\xi,\eta)\, d\eta$$

The total number of crossings (with arbitrary speed) per unit time is, therefore,

$$n_{\xi} = \int_{-\infty}^{+\infty} |\eta|\, p(\xi,\eta)\, d\eta \qquad (1.1)$$

In particular, the number of passages through zero

---

[*] Indeed, probability 0.1, for instance, means that in the course of 1 second, the event is to be expected to last 0.1 second.

is obtained as

$$n_0 = \int_{-\infty}^{+\infty} |\eta| \, p(0,\eta) \, d\eta \qquad (1.2)$$

Since, in the mean, the numbers of upward and down-ward crossings of the line $x=\xi$ will be equal, we conclude that, in the mean, the frequency for $x(t)$ to exceed a given value ξ will be $n_\xi/2$.

If x and $\dot{x}$ are independent [*], and $\dot{x}$ is normal $N(0,\tau^2)$,

$$p(\xi,\eta) = p(\xi) \, \frac{1}{\tau\sqrt{2\pi}} \, e^{-\frac{\eta^2}{2\tau^2}} \qquad (1.3)$$

it follows from (1.1) that

$$n_\xi = \sqrt{\frac{2}{\pi}} \, \tau \, p(\xi) \qquad (1.4)$$

If x, too, is normal $N(0,\sigma^2)$, then

$$n_\xi = \frac{1}{\pi} \, \frac{\tau}{\sigma} \, e^{-\frac{\xi^2}{2\sigma^2}} \qquad (1.5)$$

and, in particular,

$$n_0 = \frac{1}{\pi} \, \frac{\tau}{\sigma} = \frac{1}{\pi} \sqrt{-\frac{R''(0)}{R(0)}} \qquad (1.6)$$

---

[*] They are certainly orthogonal, cf. Eq. (I-5.8).

For wide-band noise, Fig. 4.2

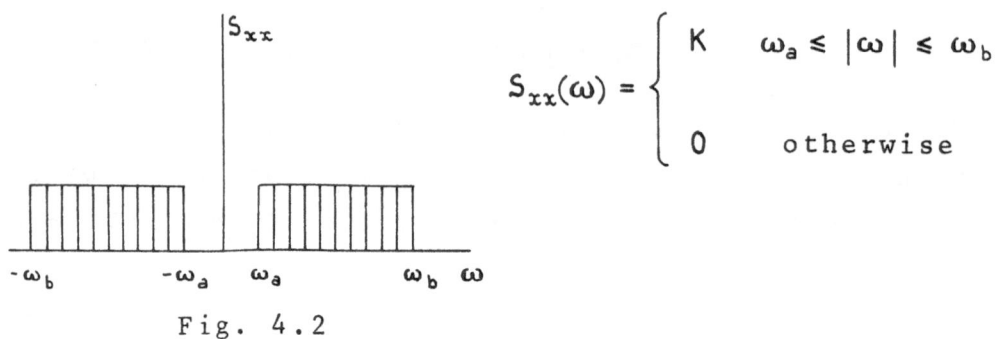

$$S_{xx}(\omega) = \begin{cases} K & \omega_a \le |\omega| \le \omega_b \\ \\ 0 & \text{otherwise} \end{cases}$$

Fig. 4.2

and we have

$$\sigma^2 = \frac{1}{\pi} K (\omega_b - \omega_a) \qquad \tau^2 = \frac{1}{\pi} \frac{K}{3} (\omega_b^3 - \omega_a^3)$$

Hence, for a normal distribution,

$$n_0 = \frac{1}{\pi} \sqrt{\frac{\omega_a^2 + \omega_a \omega_b + \omega_b^2}{3}} \qquad\qquad (1.7)$$

2. Extreme Values.

We now ask for the probability distribution of the extreme values and, in particular, of the peaks of a stochastic process, cf. [1] , [2] and [3] , p. 95.

A peak occurs if $\dot{x}(t) = 0$ and $\ddot{x}(t) < 0$. Therefore, the probability for a maximum within the intervals dt and $d\xi$ is, since $\dot{x}(t)$ has to change from + to - within that interval, given by

$$P\left[\text{Max in } d\xi\, dt\right] = P\left[\xi < x(t) < \xi + d\xi,\ \eta > \dot{x}(t) = 0 > \eta - d\eta,\ -\infty < \ddot{x} < 0\right]$$

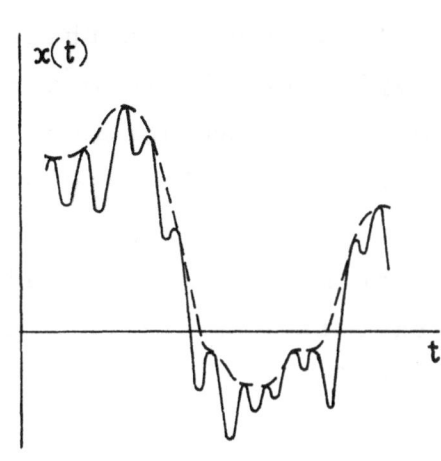

$x(t)$

Fig. 4.3

Denoting by $p(\xi, \eta, \zeta; t)$ the joint p. density of $x, \dot{x}, \ddot{x}$, we have, with $d\eta = \zeta dt$,

$$P\left[\text{Max in } d\xi\, dt\right] =$$

$$= \int_{-\infty}^{0} p(\xi, 0, \zeta; t)\, d\xi\, (-\zeta dt)\, d\zeta =$$

$$= - d\xi\, dt \int_{-\infty}^{0} \zeta p(\xi, 0, \zeta; t)\, d\zeta \qquad (2.1)$$

From this relation, the probability of a maximum within unit time above $x = a$ is

$$P\left[\text{Max} > a \text{ per unit time}\right] = -\int_{a}^{\infty} d\xi \int_{-\infty}^{0} \zeta p(\xi, 0, \zeta; t)\, d\zeta \qquad (2.2)$$

However, this probability is also equal to the expected number $N_a(t)$ of maxima above $x = a$, per unit time. Integration over a given period of time renders the expected number of maxima within that period. In the stationary case, $p(\xi, \eta, \zeta; t)$ does not depend on t .

If $x(t)$ is a stationary random process with narrow frequency band its realizations have the appearance of sine waves with slow, random amplitude - and frequency modulation, Fig. 4.4. In that case, one may

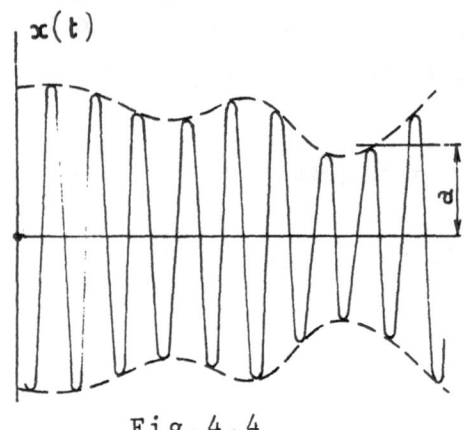

Fig.4.4

safely assume that there is only one single maximum or minimum between two zero crossings. Of n_0 crossings only n_a have, in the mean, an amplitude larger than a. Therefore, approximately,

$$P\left[\text{Max}>a \text{ per unit time}\right] = \frac{n_a}{n_0} \qquad (2.3)$$

and we have for the probability density of maxima > a

$$p_M(a) = -\frac{dP}{da} = -\frac{1}{n_0}\frac{dn_a}{da} \qquad (2.4)$$

We return now to Eq. (2.2) and assume $x(t)$ stationary, ergodic and normal with zero mean. Then

$$E\{x\dot{x}\} = 0 \;,\; E\{\dot{x}\ddot{x}\} = 0 \;,\; E\{x\ddot{x}\} = \lim \frac{1}{2T}\int_{-T}^{+T} x\ddot{x}\, dt =$$

$$= \lim \frac{1}{2T}\left[x\dot{x}\Big|_{-T}^{+T} - \int_{-T}^{+T}\dot{x}^2\, dt\right] = -E\{\dot{x}^2\} \qquad (2.5)$$

From Eq. (I-2.35), with zero mean values, one finds

$$p(x,y,z) = \frac{1}{\sqrt{8\pi^3 D}} \exp\left[-\frac{1}{2D}(m_{11}x^2 + m_{22}y^2 + m_{33}z^2 + 2m_{12}xy + 2m_{23}yz + 2m_{31}zx)\right]$$

where

$$D = \begin{vmatrix} \sigma_1^2 & \sigma_{12} & \sigma_{13} \\ \\ \sigma_{12} & \sigma_2^2 & \sigma_{23} \\ \\ \sigma_{13} & \sigma_{23} & \sigma_3^2 \end{vmatrix}, \quad m_{11} = \sigma_2^2\sigma_3^2 - \sigma_{23}^2, \quad m_{22} = \sigma_3^2\sigma_1^2 - \sigma_{31}^2,$$

$$m_{33} = \sigma_1^2\sigma_2^2 - \sigma_{12}^2$$

$$(2.6)$$

$$m_{12} = \sigma_{31}\sigma_{32} - \sigma_3^2\sigma_{12}, \quad m_{23} = \sigma_{12}\sigma_{13} - \sigma_1^2\sigma_{23}, \quad m_{31} = \sigma_{23}\sigma_{12} - \sigma_2^2\sigma_{31},$$

$$\sigma_1^2 = E\{x^2\}, \quad \sigma_{12} = E\{xy\} \quad \text{etc.}$$

With $x = \xi$, $y = 0$, $z = \zeta$ and $\sigma_{12} = \sigma_{23} = 0$, $\sigma_{13} = -\sigma_2^2$ this goes over into

$$p(\xi,0,\zeta) = \frac{1}{\sqrt{8\pi^3 D}} \exp\left[-\frac{1}{2D}(\sigma_2^2\sigma_3^2\xi^2 + 2\sigma_2^4\xi\zeta + \sigma_1^2\sigma_2^2\zeta^2)\right]$$

$$(2.7)$$

$$D = \sigma_2^2 \left(\sigma_1^2 \sigma_3^2 - \sigma_2^2 \right)$$

Differentiating Eq. (2.2), in accordance with Eq.(2.4), with respect to **a** , and substituting Eq. (2.7) one obtains, after lengthy manipulations, for the p.-density of the envelope

$$p_M(a) = \frac{1}{\sigma_1} \sqrt{\frac{1-\alpha^2}{2\pi}} \exp\left[- \frac{a^2}{2\sigma_1^2(1-\alpha^2)} \right] +$$

$$+ \frac{a}{2\sigma_1^2} \alpha \left[1 + \mathrm{erf}\left(\frac{a}{\sigma_1\sqrt{2}} \frac{\alpha}{\sqrt{1-\alpha^2}} \right) \right] \exp\left(- \frac{a^2}{2\sigma_1^2} \right) \qquad (2.8)$$

$$\alpha = \frac{n_0^+}{N} \quad , \quad \mathrm{erf}(x) = \frac{2}{\sqrt{\pi}} \int_0^x e^{-t^2} dt$$

where

$$n_0^+ = \frac{1}{2\pi} \frac{\sigma_2}{\sigma_1} \quad , \quad N = - \int_{-\infty}^{+\infty} d\xi \int_{-\infty}^{0} \zeta p(\xi,0,\zeta)\, d\zeta \qquad (2.9)$$

N is the total number of maxima per unit time, independent of their magnitude.

It can be shown that α must lie in the interval $[0,1]$. The upper limit, $\alpha = 1$, corresponds to the case of a narrow frequency band, $N = \frac{n_a}{2}$.

Eq. (2.8) reduces here to a <u>Rayleigh distribution</u>

$$p_M(a) = \frac{a}{\sigma_1^2} \exp\left(-\frac{a^2}{2\,\sigma_1^2}\right) \tag{2.10}$$

If, on the other hand, α is very small, we have the case where the mean number of maxima is much larger than that of zero crossings, cf. Fig. 4.3. Eq. (2.8) then, in the limit $\alpha = 0$, goes over into a <u>normal distribution</u>:

$$p_M(a) = \frac{1}{\sigma_1\sqrt{2\pi}} \exp\left(-\frac{a^2}{2\,\sigma_1^2}\right) \tag{2.11}$$

If $x(t)$ is not normal as, for instance, in the case of a nonlinear structure, the evaluation of Eq. (2.2) not only presents great difficulties but, moreover, there exists, in general, no possibility at all to determine $p(x, \dot{x}, \ddot{x}; t)$, even for a Markov process. The simple reason for this lies in the fact that the differential equation of a mechanical system is of order two and, thus, permits the determination of $p(x, \dot{x}, t)$ only. In order to improve on this situation it has been suggested, [1] , to employ a different definition of the concept of envelope. Let the restoring force of the system have the potential $V(x)$. If one then imagines, at every instant, the total energy of the system (per unit mass) to be transformed into potential energy,

$$\frac{\dot{x}^2}{2} + V(x) = V(a) \qquad (2.12)$$

one obtains the maximum possible amplitude a at
that instant. The curve in the x, t-diagram, formed
by these amplitudes, may also be considered an envelope whose p.-distribution is then given by

$$P\left[|x| \leqslant a\right] = \int_{-a}^{a} dx \int_{-y}^{y} p(x, \dot{x}) \, d\dot{x} = 4 \int_{0}^{a} dx \int_{0}^{y} p(x, \dot{x}) \, d\dot{x}$$

with

$$y(x, a) = \sqrt{2 \left[V(a) - V(x)\right]} \qquad (2.13)$$

Differentiation renders the p.-density of the envelope as

$$p_M(a) = \frac{dP}{da} = 4V'(a) \int_{0}^{a} \frac{p\left[x, y(x, a)\right]}{y(x, a)} \, dx \qquad (2.14)$$

We observe that now only knowledge of $p(x, \dot{x})$ is ne
cessary for the determination of $p_M(a)$.

3. Criteria of Fatigue.

          In order to obtain some measure of material
fatigue in the case of a randomly fluctuating load

the <u>Palmgren-Miner</u> law will be assumed to be valid.
According to this law, damage produced by the i-th
load cycle is equal to, [1] ,

$$s_i = k a_i^\alpha$$

where a_i is the stress amplitude, and k and α are
material constants. The damages caused by subsequent
load cycles are assumed to superpose linearly. Hence,
after n cycles damage has increased to

$$S = \sum_{i=1}^{n} s_i$$

Now, if a_i is a random variable, S too
is random, and we find for the expectation of damage
after n cycles

$$E\{S\} = \sum E\{s_i\} = nk\, E\{a^\alpha\} = nk \int_0^\infty a^\alpha\, p_M(a)\, da \qquad (3.1)$$

If we refer the process to the time interval $(0,T]$ rather than to the number n of load
cycles, we may write, in the stationary case,

$$n = n_0^+ T = \frac{n_0}{2} T \qquad (3.2)$$

assuming a narrow frequency band, Fig. 4.4.

In order to make the diagrams and tables obtained in the usual deterministic fatigue tests of materials applicable to random load fluctuations, J.W. Miles has suggested, [7] , to introduce an "equivalent" cycle stress, whose frequency is n_0^+, and whose amplitude is given by

$$A^\alpha = \int_0^\infty a^\alpha \, p_M(a) \, da \qquad (3.3)$$

The total damage is then the same in both cases[*].

We also mention another mathematical model frequently used for the determination of the expected lifetime of structural, in particular electronic, components (radio tubes, for instance). The model is characterized by the two following assumptions, see [4], p. 61:

(a) If the lifetime of the component has reached the value t , the probability that it will fail in the time interval $(t, t + dt]$ is given by

$$P_+ [\text{failure in the interval } dt] = a(t) \, dt$$

where $a(t)$ is a given function, compensating for the increasing wear and, hence, increasing probability of failure with increasing service time of the component.

---

[*] For further details refer to [6] and [8].

(b) The probability of putting a defective component into service is zero:

$$P_t = 0 \qquad \text{for} \qquad t \leq 0$$

Let $F(t)$ denote the p.-distribution of the component

$$P[T \leq t] = F(t)$$

where T represents lifetime. Now

$$P[t < T \leq t + dt] = P[T \leq t + dt \mid T > t] \, P[T > t]$$

Since

$$P[t < T \leq t + dt] = F(t + dt) - F(t) = F'(t) dt + \ldots$$

$$P[T > t] = 1 - F(t)$$

and, according to (a),

$$P[T \leq t + dt \mid T > t] = a(t) \, dt$$

we obtain the differential equation

$$F'(t) = [1 - F(t)] \, a(t)$$

Integration renders, with initial condition taken into consideration,

$$F(t) = 1 - \exp\left(-\int_0^t a(\tau)\, d\tau\right) \quad (t \geq 0) \qquad (3.4)$$

A frequently used expression for the function $a(t)$ is

$$a(t) = \alpha + \beta e^{\gamma t} \qquad (3.5)$$

Putting $\alpha = 0$ one obtains the Weibull-distribution[*].

Literature .

[1] S.H. Crandall: Zero crossings, peaks, and other
    statistical measures of random responses.
    J. Acoust. Soc. Am. 35 (1963), 1693

[2] Y.K. Lin: Probability distributions of stress
    peaks in linear and nonlinear structures.
    AIAA-J. 1(1963), 1133.

[3] J.D. Robson: Random Vibration. Elsevier.Amsterdam
    1964.

[4] J. Heinhold and K.W. Gaede: Ingenieur-Statistik.
    Oldenbourg-Verlag. München-Wien 1964.

[5] M.Shinozuka and J.T.P. Yao: On the two-sided time
    dependent barrier problem, J.Sound Vib. 6 (1967)
    98.

[6] R.W.Lardner: A theory of random fatigue. J.Mech.
    Phys. Solids 15(1967),205.

[7] J.W.Miles: On structural fatigue under random
    loading. J.Aeron. Sci. 21 (1954), 753.

[8] S.H. Crandall (editor): Random Vibration. Vol.2,
    p.126 ff. M.I.T. Press 1963.

[9] A.M. Freudenthal: Combination of the theories
     of elasticity, plasticity and viscosity in study
     ing the safety of structures. Eighth Congress
     Int. Ass. Bridge Struct. Enging., New York
     1968.

CHAPTER V

STOCHASTIC STABILITY .

We turn now to the investigation of differ‑
ential equations whose coefficients are random func‑
tions ("parametric excitation"). Such equations occur
frequently in applications: flutter of aircraft wings
in turbulent atmosphere, instruments on shaking ground
or shaking suspensions, wave propagation in inhomoge‑
neous media etc. In this connection, the question of
stability or instability of the motion is of fundamen‑
tal importance .

We first give a simple example.

1. Pendulum with Randomly Shaken Suspension Point .

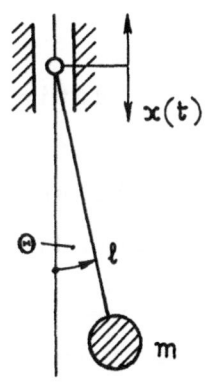

Let the suspension point of
an oscillating pendulum move verti‑
cally under the action of random
forces, Fig. 5.1. The law of angu‑
lar momentum, referred to the sus‑
pension point, then renders

Fig. 5.1

$$m\ell^2\ddot{\Theta} - m\ell\ddot{x}\sin\Theta = -mg\ell\sin\Theta - \lambda\dot{\Theta}$$

where λ is the coefficient of damping. For small amplitudes this simplifies to

$$\ddot{\Theta} + 2a\,\dot{\Theta} + \left[\omega_0^2 - \alpha(t)\right]\Theta = 0 \qquad (1.1)$$

where

$$2a = \frac{\lambda}{m\,\ell^2} \quad, \quad \omega_0^2 = \frac{g}{\ell} \quad, \quad \alpha(t) = \frac{\ddot{x}(t)}{\ell}$$

An extensive literature[*] exists for differ̲ ential equation (1.1) and related equations for the case where $\alpha(t)$ is a deterministic, periodic func- tion. If, however, $\alpha(t)$ is random, the corresponding motion is random and the question of its stability becomes a probabilistic problem. To solve it the con- cept of stability must first be generalized to random functions.

2. Definition of Stochastic Stability .

We first recall the deterministic defini- tion of Lyapunov stability.

Let $x \equiv 0$ be the equilibrium or null solu- tion whose stability properties are being tested.

---

[*] See, for instance, $\begin{bmatrix} 1 \end{bmatrix}$, where a rather complete list of references may be found.

$x = (x_1, x_2 \ldots x_n)$ is assumed to be an n-vector. Now let the equilibrium be disturbed and the system be moved to an initial state x_0 at time $t = t_0$. Denote the ensuing solution by $x(t)$. The equilibrium solution is then said to be stable in the sense of Lyapunov if, given $\varepsilon > 0$, there exists an $\eta(\varepsilon, t_0)$ such that, for all $t \geq t_0$,

$$\| x(t) \| < \varepsilon \qquad (2.1)$$

provided $\| x_0 \| < \eta$ initially. $\| x(t) \|$ denotes the norm

$$\| x(t) \| = \sum_{i=1}^{n} | x_i(t) | \qquad (2.2)$$

If η can be chosen independently of t_0, the equilibrium solution $x \equiv 0$ is said to be <u>uniformly stable</u>.

If $x \equiv 0$ is stable and if, in addition,

$$\lim_{t \to \infty} \| x(t) \| = 0 \qquad (2.3)$$

equilibrium is said to be <u>asymptotically stable</u>. Finally, if (2.3) holds for any x_0, the equilibrium solution is said to be asymptotically <u>stable in the large</u>.

Turning to the concept of stochastic stability, we note that $\| x(t) \|$ is now a random variable. Hence, definitions (2.1) and (2.3) must be reformulated, which can be done in various ways, cf. [2] . Three widely used definitions are listed in the following.

(a) Stability of the Mean. The equilibrium solution has mean stability if, given $\varepsilon > 0$, there exists $\eta(\varepsilon, t_0)$ such that, for all $t \geqslant t_0$,

$$E\left\{\|x(t)\|\right\} < \varepsilon \qquad (2.4)$$

provided $\|x_0\| < \eta$ initially.

If

$$\lim_{t \to \infty} E\left\{\|x(t)\|\right\} = 0 \qquad (2.5)$$

we have asymptotic stability of the mean.

(b) Stability of Mean Square. The equilibrium solution has mean square stability if, given $\varepsilon > 0$, there exists $\eta(\varepsilon, t_0)$ such that, for all $t \geqslant t_0$,

$$E\left\{\|x(t)\|_2\right\} < \varepsilon \qquad (2.6)$$

provided $\|x_0\| < \eta$ initially. $\|x(t)\|_m$ is defined by

$$\|x\|_m = \sum_{i=1}^{n} |x_i|^m \qquad (2.7)$$

If

$$\lim_{t \to \infty} E \left\{ \| x(t) \|_2 \right\} = 0 \qquad (2.8)$$

the equilibrium solution is said to have asymptotic stability of the mean square.

(c) <u>Almost Sure Asymptotic Stability</u>. The equilibrium solution is said to be almost surely asymptotically stable (or asymptotically stable with probability one) if, for any $\varepsilon > 0$,

$$\lim_{t \to \infty} P \left[\| x(t) \| > \varepsilon \right] = 0 \qquad (2.9)$$

Stronger stability definitions are obtained if $\| x(t) \|$ is replaced by $\sup_{t \ge t_0} \| x(t) \|$. The study of the latter random variable, however, requires much more sophisticated techniques.

3. Sufficient Conditions for Almost Sure Stability.

Consider the <u>linear</u> system of differential equations

$$\dot{x}_i = \sum_j \left[a_{ij} + f_{ij}(t) \right] x_j \qquad (i, j = 1, 2 \ldots n) \quad (3.1)$$

where the $f_{ij}(t)$ are random processes, stationary in

the strict sense, and continuous and ergodic with
probability one. From continuity, existence and uni-
queness with probability one of the solution of (3.1)
follow in $[0,\infty)$.

Kozin [3] was the first to give a sufficient
condition for almost sure stability. For its derivation we need the Gronwall-Bellman lemma: If

$$u(t) \leq C + \int_0^t u(\tau)v(\tau)\,d\tau \qquad (a)$$

with $u(t) \geq 0$, $v(t) \geq 0$, and $C \geq 0$,   then, also

$$u(t) \leq C \exp \int_0^t v(\tau)\,d\tau \qquad (b)$$

Proof: From (a)

$$\frac{uv}{C + \int_0^t uv\,d\tau} \leq v$$

and, upon integration,

$$\log\left(C + \int_0^t uv\,d\tau\right) - \log C \leq \int_0^t v\,d\tau$$

where the constant of integration C has been put
equal to zero which, obviously, represents the strong
est condition.

Therefore,

$$C + \int_0^t u \upsilon \, d\tau \leq C \exp \int_0^t \upsilon \, d\tau, \qquad \text{Q.E.D.}$$

We now write the solution of Eq. (3.1) in the form

$$x_i(t) = y_i(t) + \sum_j \sum_k \int_0^t Y_{ij}(t-\tau) f_{jk}(\tau) x_k(\tau) \, d\tau \qquad (3.2)$$

where $y_i(t)$ is that solution of the system

$$\dot{x}_i = \sum_j a_{ij} x_j \qquad (3.3)$$

for which initially $y_i(0) = x_i(0)$, while $Y_{ij}(t)$ represents the fundamental system of solutions of (3.3) for which initially $Y_{ij}(0) = \delta_{ij}$, cf. (II-2.6). Using now the norm (2.2) for the vector $\underset{\sim}{x}(t) = [x_1(t), \ldots x_n(t)]$, and the norm

$$\| M \| = \sum_i \sum_j |m_{ij}|$$

for the matrix $M = (m_{ij})$, we obtain from Eq. (3.2) and as a consequence of

$$\| a + b \| \leq \| a \| + \| b \|, \quad \| M \underset{\sim}{x} \| \leq \| M \| \cdot \| x \|$$

the inequality

$$\| x(t) \| \leq \| y(t) \| + \int_0^t \| Y(t-\tau) \| \cdot \| F(\tau) \| \cdot \| x(\tau) \| \, d\tau \qquad (3.4)$$

Since we have assumed asymptotic stability (sufficient conditions), we may write

$$\| y(t) \| \le a e^{-\alpha t}, \quad \| Y(t) \| \le b e^{-\alpha t} \qquad (3.5)$$

with suitably chosen positive a, b and α. Then, from Eq. (3.4),

$$\| x(t) \| \le e^{-\alpha t} \left[a + b \int_0^t e^{\alpha \tau} \| F(\tau) \| \cdot \| x(\tau) \| d\tau \right]$$

Applying the Gronwall-Bellman lemma we obtain

$$\| x(t) \| \le a \exp \left[-\alpha t + b \int_0^t \| F(\tau) \| d\tau \right]$$

If the exponent on the right-hand side is written in the form

$$\left[-\alpha + b \frac{1}{t} \int_0^t \| F(\tau) \| d\tau \right] t$$

it follows, as a consequence of the ergodicity of the $f_{ij}(t)$, i.e., or

$$\lim_{t \to \infty} \frac{1}{t} \int_0^t \| F(\tau) \| d\tau = E \left\{ \| F(\tau) \| \right\}$$

that the condition

$$E \left\{ \| F(t) \| \right\} < \frac{\alpha}{b} \qquad (3.6)$$

is sufficient for almost sure stability.

Condition (3.6) severely restricts the class of "stable functions" $f_{ij}(t)$ and is, therefore of little practical value. Infante [4] has given a considerably improved condition. For its derivation, we rewrite Eq. (3.1) in matrix form

$$\dot{x}(t) = \left[A + F(t) \right] x(t) \qquad (3.1a)$$

and assume, in addition,

$$E\left\{ F(t) \right\} = 0 \qquad (3.7)$$

We need the following lemma whose proof may be found in the textbooks on matrix theory:

Lemma. Let B and D be two real, symmetric $n \times n$ matrices, with B positive definite. Then the matrix DB^{-1} possesses n real eigenvalues λ_i, and

$$\lambda_{min} = \min_{x} \frac{x^T D x}{x^T B x} \quad , \quad \lambda_{max} = \max_{x} \frac{x^T D x}{x^T B x}$$

Consider now the quadratic Lyapunov function

$$V(x) = x^T B x$$

Since B is positive definite, so is $V(x)$. Different-iation renders

$$\dot{V} = \dot{x}^T B x + x^T B \dot{x}$$

and we have, along a solution of (3.1a)

$$\dot{V} = \left[(A+F)x\right]^T B x + x^T B (A+F) x = x^T (A+F)^T B x + x^T B (A+F) x$$

Now we form

$$\lambda(t) = \frac{\dot{V}(x)}{V(x)} = \frac{x^T\left[(A+F)^T B + B(A+F)\right]x}{x^T B x}$$

Putting

$$(A+F)^T B + B(A+F) = D$$

the lemma may be employed, and $\lambda(t)$ may be bounded from below and above as

$$\lambda_{min}\left(DB^{-1}\right) \leq \lambda(t) \leq \lambda_{max}(DB^{-1})$$

According to Lyapunov's theory, the motion (3.1a) is asymptotically stable for $\dot{V}$ negative. Hence, in the present case,

$$V = V\left[x(0)\right] \exp \int_0^t \lambda(\tau)\,d\tau \equiv V\left[x(0)\right] \exp \left[\frac{1}{t}\int_0^t \lambda(\tau)\,d\tau\right]t$$

must be bounded and approach zero with probability one for $t \longrightarrow \infty$. Again, the ergodic theorem renders

$$E\left\{\lambda(t)\right\} < 0$$

and this leads to the theorem of Infante:

Theorem: If, for a positive definite matrix B and some $\varepsilon > 0$,

$$E\left\{\lambda_{max}\right\} \leq - \varepsilon$$

where λ_{max} is the largest eigenvalue of the matrix

$$A^T + F^T + B(A + F) B^{-1} \tag{3.8}$$

the system (3.1a) is almost surely asymptotically stable in the large.

The theorem gives no clue as to the construction of the optimal matrix B . Furthermore, since the matrix (3.8) contains the matrix $F(t) \equiv \left(f_{ij}(t)\right)$ which is only statistically known, the theorem is difficult to apply. Infante has, therefore, given two corollaries which are easier to handle but place more severe restrictions on the functions $f_{ij}(t)$. We list the second without proof:

Corollary. if the matrix $F(t)$ in Eq. (3.1a) is written in the form

$$F(t) = \sum_{i=1}^{m} \varphi_i(t) C^{(i)} \quad \text{or} \quad f_{ij}(t) = \sum_{k=1}^{m} \varphi_k(t) C_{ij}^{(k)} \tag{3.9}$$

with $m \leq n^2$, and if, with a positive definite ma-

---

trix **B**   and some $\varepsilon > 0$,

$$\frac{1}{2} \sum_{i=1}^{m} \left(\mu_{max}^{(i)} - \mu_{min}^{(i)} \right) E \left\{ |\varphi_i(t)| \right\} + \nu_{max} \leq -\varepsilon \qquad (3.10)$$

then (3.1a) is almost surely asymptotically stable
in the large. The $\mu^{(i)}$ are the eigenvalues of the
matrix

$$C^{(i)T} + BC^{(i)} B^{-1}$$

and ν_{max} is the largest eigenvalue of the matrix

$$A^T + BAB^{-1}$$

We note that, now, only the eigenvalues of
known matrices have to be calculated.

As an **example**, consider the equation of
the linear oscillator with parametric excitation

$$\ddot{x} + 2\zeta\dot{x} + \left[1 + f(t) \right] x = 0 \qquad (3.11)$$

Putting $x = x_1$, $\dot{x} = x_2$ we have

$$A = \begin{pmatrix} 0 & 1 \\ -1 & -2\zeta \end{pmatrix}, \quad C = \begin{pmatrix} 0 & 0 \\ -1 & 0 \end{pmatrix} \qquad (3.12)$$

We first apply **Kozin**'s theorem. From the
general solution of (3.11), with $f(t) \equiv 0$, one has

at once for the bounds in (3.5)

$$\alpha = \xi \; , \qquad b = \frac{4}{\sqrt{1-\xi^2}}$$

Since

$$\| F(t) \| = | f(t) |$$

Eq. (3.6) renders

$$E \left\{ | f(t) | \right\} < \frac{\xi}{4} \sqrt{1 - \xi^2} \qquad\qquad (3.13)$$

Now we use <u>Infante</u>'s theorem and choose for the matrix B the most general quadratic, positive definite form

$$B = \begin{pmatrix} \alpha_1^2 + \alpha_2 & \alpha_1 \\ \alpha_1 & 1 \end{pmatrix} \; , \; \alpha_2 > 0 \qquad (3.14)$$

with α_1 and α_2 as yet undetermined. They will be used later to make B optimal.

First we have

$$B^{-1} = \frac{1}{\alpha_2} \begin{pmatrix} 1 & -\alpha_1 \\ -\alpha_1 & \alpha_1^2 + \alpha_2 \end{pmatrix}$$

and, therefore, for the matrix $B \left(A + F \right) B^{-1}$

$$\frac{1}{\alpha_2} \begin{pmatrix} -\alpha_1(1+f)-\alpha_1^2(\alpha_1-2\zeta)-\alpha_1\alpha_2 & \alpha_1^2(1+f)+(\alpha_1^2+\alpha_2)\left[\alpha_1(\alpha_1-2\zeta)+\alpha_2\right] \\ \\ -(1+f)-\alpha_1(\alpha_1-2\zeta) & \alpha_1(1+f)+(\alpha_1-2\zeta)(\alpha_1^2+\alpha_2) \end{pmatrix}$$

The largest eigenvalue of the matrix (3.8) follows as

$$\lambda_{max} = -2\zeta + \sqrt{4(\zeta-\alpha_1)^2 + \frac{1}{\alpha_2}\left[\alpha_2+\alpha_1^2-1-f(t)+2\alpha_1(\zeta-\alpha_1)\right]^2} \qquad (3.15)$$

Putting $f = 0$ we obtain ν_{max} . Finally

$$C^T + BCB^{-1} = \frac{-1}{\alpha_2}\begin{pmatrix} \alpha_1 & \alpha_2-\alpha_1^2 \\ \\ 1 & -\alpha_1 \end{pmatrix}$$

whence

$$\mu_{max} = +\frac{1}{\sqrt{\alpha_2}} \quad , \quad \mu_{min} = -\frac{1}{\sqrt{\alpha_2}}$$

The theorem then renders

$$E\left\{-2\zeta + \sqrt{4(\zeta-\alpha_1)^2 + \frac{1}{\alpha_2}\left[\alpha_2+\alpha_1^2-1-f(t)+2\alpha_1(\zeta-\alpha_1)\right]^2}\right\} < 0$$

and the corollary gives

$$\frac{1}{\sqrt{\alpha_2}} \; E\left\{|f(t)|\right\} < 2\zeta - \sqrt{4(\zeta-\alpha_1)^2 + \frac{1}{\alpha_2}\left[\alpha_2+\alpha_1^2-1+2\alpha_1(\zeta-\alpha_1)\right]^2}$$

We now choose α_1 and α_2 such as to make

$E\{|f(t)|\}$ =max. First, from

$$\frac{\partial}{\partial \alpha_1} E\{.\} = 0$$

$\alpha_1 = \xi$ follows. Upon substitution α_2 follows by inspection, and one finds

$$\alpha_1 = \xi \quad, \quad \alpha_2 = 1 - \xi^2, \text{ if } \quad \xi \leq 1/\sqrt{2}$$

$$\alpha_1 = \xi \quad, \quad \alpha_2 = \xi^2, \quad \text{ if } \quad \xi \geq 1/\sqrt{2}$$

and, hence, from the theorem,

$$E\{|f(t)|\} < 2\xi \sqrt{1-\xi^2}, \qquad \xi \leq 1/\sqrt{2}$$

$$E\{|f(t)+1-2\xi^2|\} < 2\xi^2, \qquad \xi \geq 1/\sqrt{2}$$

$$(3.16)$$

while, from the corollary,

$$E\{|f(t)|\} < 2\xi \sqrt{1-\xi^2} \quad, \qquad \xi \leq 1/\sqrt{2}$$

$$E\{|f(t)|\} \leq 1 \quad, \qquad\qquad \xi \geq 1/\sqrt{2}$$

$$(3.17)$$

One notes that, for $\xi \leq 1/\sqrt{2}$, the same bounds are rendered by theorem and corollary, while, for strong damping, $\xi > 1/\sqrt{2}$, the conditions of the theorem are considerably weaker. In addition, a comparison with (3.13) shows that the bound obtained from Kozin's theorem is much lower and, in fact, practically useless. This becomes particularly obvious if one transforms from $E\{|f(t)|\}$ to $E\{f^2(t)\}$ with the aid of the Schwarz inequality

$$E^2\{|f(t)|\} \leqslant E\{f^2(t)\}$$

4. Stability of the Mean in the Presence of White Noise.

Frequently, one is satisfied with investigating stability of the mean or, eventually, mean square. The investigation of the stability of higher moments leads, in general, already to considerable computational difficulties.

In addition to restricting ourselves to linear systems we assume the coefficients to be white noise. It is not difficult to develop a theory valid for a system of n differential equations [5] . However, here we consider only one equation:

$$\ddot{y} + [a + \alpha_1(t)]\dot{y} + [\omega_0^2 + \alpha_0(t)]y = f(t) + \alpha_2(t) \qquad (4.1)$$

where a and ω_0^2 are constants, $f(t)$ is a deterministic excitation, and $\alpha_0(t)$, $\alpha_1(t)$ and $\alpha_2(t)$ are stationary, normal, white random processes with

$$< \alpha_i(t) > = 0 , \quad < \alpha_i(t)\alpha_j(s) > = 2 D_{ij}\delta(t-s) \qquad (4.2)$$

where $D_{ij} = D_{ji}$.

The Fokker-Planck equation (I-9.20), corresponding to (4.1), is

$$\frac{\partial p}{\partial t} = - \frac{\partial(\dot{y}p)}{\partial y} - \frac{\partial}{\partial \dot{y}} \left\{ \left[F(t) - D_{21} - (\omega_0^2 - D_{10})\, y - (a - D_{11})\, \dot{y} \right] p \right\} +$$

$$+ \frac{\partial^2}{\partial \dot{y}^2} \left\{ \left[D_{00}\, y^2 + (D_{10} + D_{01})\, y\dot{y} + D_{11}\, \dot{y}^2 - \right. \right. \qquad (4.3)$$

$$\left. \left. - 2\,(D_{20}\, y + D_{21}\, \dot{y}) + D_{22} \right] p \right\} = 0$$

For the expectation of a function $F(y, \dot{y})$,

$$< F(y, \dot{y}) > = \int\limits_{-\infty}^{+\infty} \int\limits_{-\infty}^{+\infty} F(y, \dot{y})\, p(y, \dot{y}\,;\, t)\, dy\, d\dot{y} \qquad (4.4)$$

one obtains, upon multiplying (4.3) by F, and integrating,

$$\frac{d<F>}{dt} = - \int\limits_{-\infty}^{+\infty} \left[\dot{y}\, Fp\, \Big|_{y=-\infty}^{+\infty} - \int\limits_{-\infty}^{+\infty} \dot{y}\, \frac{\partial F}{\partial y}\, p\, dy \right] d\dot{y} - \left[F(t) - D_{21} \right] \int\limits_{-\infty}^{+\infty} \left[Fp\, \Big|_{\dot{y}=-\infty}^{+\infty} - \int\limits_{-\infty}^{+\infty} \frac{\partial F}{\partial \dot{y}}\, p\, d\dot{y} \right] dy +$$

$$+ \left(\omega_0^2 - D_{10} \right) \int\limits_{-\infty}^{+\infty} \left[Fpy\, \Big|_{\dot{y}=-\infty}^{+\infty} - \int\limits_{-\infty}^{+\infty} y\, \frac{\partial F}{\partial \dot{y}}\, p\, d\dot{y} \right] dy + (a - D_{11}) \left[- \int\limits_{-\infty}^{+\infty}\!\!\int \frac{\partial F}{\partial \dot{y}}\, \dot{y}p\, dy\, d\dot{y} \right] +$$

$$+ D_{00} \left[-\iint\limits_{-\infty}^{+\infty} \frac{\partial}{\partial y}(y^2 F) \frac{\partial p}{\partial \dot{y}} \, dy \, d\dot{y} \right] + 2 D_{10} \left[-\iint\limits_{-\infty}^{+\infty} \frac{\partial}{\partial \dot{y}} (yF) \frac{\partial(\dot{y}p)}{d\dot{y}} \, dy \, d\dot{y} \right] +$$

$$+ D_{11} \left[-\iint\limits_{-\infty}^{+\infty} \frac{\partial F}{\partial y} \frac{\partial(\dot{y}^2 p)}{\partial \dot{y}} \, dy \, d\dot{y} \right] + 2 D_{20} \iint\limits_{-\infty}^{+\infty} \frac{\partial}{\partial \dot{y}} (yF) \frac{\partial p}{\partial \dot{y}} \, dy \, d\dot{y} +$$

$$+ 2 D_{21} \iint\limits_{-\infty}^{+\infty} \frac{\partial F}{\partial \dot{y}} \frac{\partial(\dot{y}p)}{\partial \dot{y}} \, dy \, d\dot{y} + 2 D_{22} \iint\limits_{-\infty}^{+\infty} \frac{\partial F}{\partial \dot{y}} \frac{\partial p}{\partial \dot{y}} \, dy \, d\dot{y}$$

Here, integration by parts has been used. The integrated parts vanish (they have been written out explicitly only for the first three terms), and one ob
tains, after a second integration by parts,

$$\frac{d\langle F \rangle}{dt} = \langle \dot{y} \frac{\partial F}{\partial y} \rangle + \left[f(t) - D_{21} \right] \langle \frac{\partial F}{\partial \dot{y}} \rangle - (\omega_0^2 - D_{10}) \langle y \frac{\partial F}{\partial \dot{y}} \rangle -$$

$$- (a - D_{11}) \langle \dot{y} \frac{\partial F}{\partial \dot{y}} \rangle + D_{00} \langle y^2 \frac{\partial^2 F}{\partial \dot{y}^2} \rangle + 2 D_{10} \langle y\dot{y} \frac{\partial^2 F}{\partial \dot{y}^2} \rangle +$$

$$+ D_{11} \langle \dot{y}^2 \frac{\partial^2 F}{\partial \dot{y}^2} \rangle - 2 \left[D_{20} \langle y \frac{\partial^2 F}{\partial \dot{y}^2} \rangle + D_{21} \langle \dot{y} \frac{\partial^2 F}{\partial \dot{y}^2} \rangle + D_{22} \langle \frac{\partial^2 F}{\partial \dot{y}^2} \rangle \right]$$

$$(4.5)$$

Putting now $F = y$ and $F = \dot{y}$, one gets, respectively,

$$\frac{d\langle y \rangle}{dt} = \langle \dot{y} \rangle \ , \quad \frac{d\langle \dot{y} \rangle}{dt} = F(t) - D_{21} - (\omega_0^2 - D_{10})\langle y \rangle - (a - D_{11})\langle \dot{y} \rangle$$

Elimination of $\langle \dot{y} \rangle$ leads to the following differential equation for the mean $\langle y \rangle$:

$$\frac{d^2\langle y \rangle}{dt^2} + (a - D_{11})\frac{d\langle y \rangle}{dt} + (\omega_0^2 - D_{10})\langle y \rangle = F(t) - D_{21} \qquad (4.6)$$

Stability of this deterministic equation may now be discussed in the usual manner, using, for instance, the Routh-Hurwitz criteria.

5. Stability of Mean Square in the Presence of White Noise.

For the sake of simplicity, we restrict ourselves to the case $\alpha_1(t) = \alpha_2(t) = 0$ in Eq. (4.1). Only the coefficient of y is then random[*] and Eq. (4.5) reduces to

$$\frac{d\langle F \rangle}{dt} = \langle \dot{y} \frac{\partial F}{\partial y} \rangle - \omega_0^2 \langle y \frac{\partial F}{\partial \dot{y}} \rangle - a\langle \dot{y} \frac{\partial F}{\partial \dot{y}} \rangle + D_{00}\langle y^2 \frac{\partial^2 F}{\partial \dot{y}^2} \rangle + F(t)\langle \frac{\partial F}{\partial \dot{y}} \rangle$$

$$(5.1)$$

---

[*] The two Fokker-Planck equations (I -9.16) and (I -9.20) become identical in this case!

By suitable choices of F one obtains

$$\frac{d\langle y\rangle}{dt}=\langle\dot{y}\rangle \; , \quad \frac{d\langle\dot{y}\rangle}{dt}=-\omega_0^2\langle y\rangle-a\langle\dot{y}\rangle+f(t) \qquad (5.2)$$

$$\left.\begin{array}{l}\dfrac{d\langle y^2\rangle}{dt}=2\langle y\dot{y}\rangle \; , \quad \dfrac{d\langle y\dot{y}\rangle}{dt}=\langle\dot{y}^2\rangle-\omega_0^2\langle y^2\rangle-a\langle y\dot{y}\rangle+f(t)\langle y\rangle \\[3mm] \dfrac{d\langle\dot{y}^2\rangle}{dt}=-2\omega_0^2\langle y\dot{y}\rangle-2a\langle\dot{y}^2\rangle+2\,D_{00}\langle y^2\rangle+2f(t)\langle\dot{y}\rangle\end{array}\right\}(5.3)$$

One observes that, for $f(t)\equiv 0$, the equations
for the various moments are decoupled: (5.2) contains
only the moments of order one, while (5.3) contains on
ly the second-order moments. If $f(t)\neq 0$, Eq. (5.3)
contains, in addition, the preceding first-order moments, but no higher ones. This is no longer so if the
coefficient $\alpha_0(t)$ is not white noise but is obtained
from white noise by linear filtering, i.e., for instance, by

$$\dot{\alpha}_0+\beta\alpha_0=\gamma w(t)$$

The equations for the moments are then coupled, $[7]$,
i.e., those of order two contain the moments of order
two and three which, in turn, depend on those of order four, etc.[*]. Obviously, therefore, it does not

---

[*] cf. section 6.

longer seem to be justified to speak then of stability of mean square.

As an **example** we consider the pendulum discussed in sec. 1. Let

$$< \alpha(t) > = 0 \; , \quad < \alpha(t) \cdot \alpha(s) > = 2 D \delta (t-s)$$

From Eqs. (5.2) one finds $a > 0$ as necessary and sufficient condition for asymptotic stability of the mean, i.e., for

$$\lim_{t \to \infty} < \Theta > \; = \; \lim_{t \to \infty} < \dot{\Theta} > = 0$$

The condition is independent of the stochastic excitation, and is satisfied only if the motion is damped.

Eqs. (5.3) represent three homogeneous equations for the three second-order moments $< \Theta^2 >$, $< \Theta \dot{\Theta} >$ and $< \dot{\Theta}^2 >$. Therefore, for asymptotic stability of the mean square, the eigenvalues of the coefficient matrix must all have negative real parts. This is the case if and only if

$$a > 0 \quad \text{and} \quad a \omega_0^2 > D$$

Damping is, therefore, necessary whose minimum value is determined by the second condition, depending on the intensity of the motion of the suspension point.

6. Linear Stochastic Equations with Non-White Parameters.

We consider a stochastic equation of the form

$$Lu(t) = g(t) \qquad\qquad (6.1)$$

where $g(t)$ is a given deterministic function, and L denotes a stochastic operator, linear in u and its derivatives. We assume L to depend on a random parameter $z(t)$, with probability density $p(z)$. Hence, $u(t)$ is a random function.

We shall attempt to obtain differential equations for the various moments of u, in particular for the expectation $< u >$. In doing so we will note, however, that the equation for the moment of one particular order will always contain moments of higher order too. One is, therefore, facing an infinite system of equations, known as "hierarchy equations".

We restrict the operator L to the following special form

$$\left[L_0 + z(t)\right] u(t) = g(t) \qquad\qquad (6.2)$$

where L_0 is a deterministic linear operator, and $z(t)$ is a random function. For $z(t)$ white noise we have then Eq. (4.1), with $\alpha_1 = \alpha_2 = 0$. Taking expectation on both sides of Eq. (6.2) we get

$$L_0 <u> + <z(t)u(t)> = g(t) \qquad (6.3)$$

Since this equation contains $<z(t)u(t)>$ as additional unknown one would, perhaps, be inclined to multiply (6.2) by $z(t)$ and average. This, however, would lead to $<z L_0 u>$, and not to $L_0<zu>$. We multiply, therefore, by $z(s)$, rather than $z(t)$, and obtain, with $z(s) L_0 z(t) = L_0 z(s) z(t)$,

$$L_0 <z(s)u(t)> + <z(s)z(t)u(t)> = <z(s)>g(t) \qquad (6.4)$$

After $<z(s)u(t)>$ has been determined from this equation, we could put $s = t$ and substitute into (6.3) to obtain an equation for $<u>$. We note, however, that (6.4) contains a new unknown $<z(s)z(t)u(t)>$. Therefore, we need an additional equation which in turn however, contains new moments, and so on ad infinitum.

One possibility to overcome this difficulty consists in the introduction of a moment-generating functional $F\{x, y\}$, see [8],

$$F\{x,y\} = <\exp \int [x(\omega)u(\omega) + y(\omega)z(\omega)] d\omega> =$$

$$= \int_{-\infty}^{+\infty}\int_{-\infty}^{+\infty} \left[\exp \int (x(\omega)u(\omega) + y(\omega)z(\omega)) d\omega\right] p(u,z;t) du\, dz \qquad (6.5)$$

whose functional derivatives[*)]

$$\frac{\delta^{i+k} F}{\delta x(t_1) \ldots \delta x(t_j) \delta y(t_{j+1}) \ldots \delta y(t_{j+k})} = <u(t_1) \ldots u(t_j) z(t_{j+1}) \ldots z(t_{j+k}) \exp\left[\int (xu + yz) d\omega\right]>$$

render, for $x = y = 0$, the moments

$$\left.\frac{\delta^{i+k} F}{\delta x(t_1) \ldots \delta x(t_j) \delta y(t_{j+1}) \ldots \delta y(t_{j+k})}\right|_{x=y=0} = <u(t_1) \ldots u(t_j) z(t_{j+1}) \ldots z(t_{j+k})>$$

As a consequence of

$$L_0 \frac{\delta F}{\delta x(t)} + \frac{\delta^2 F}{\delta x(t) \delta y(t)} = <\left[L_0 u(t) + u(t) z(t)\right] \exp\left[\int (xu + yz) d\omega\right]> =$$

$$= < g(t) \exp\left[\int (xu + yz) d\omega\right]> = g(t) < \exp \int (xu + yz) d\omega >$$

one obtains for F the following functional differen
tial equation

---

*) The Fréchet differential δF of a functional
   $F\{x(t)\}$ is defined as

$$\delta F = \frac{\partial}{\partial \varepsilon} F\{x(t) + \varepsilon \xi(t)\}\Big|_{\varepsilon = 0} = \int \frac{\delta F}{\delta x(t)} \xi(t) dt$$

whence the functional derivative $\delta F / \delta x(t)$ may be found.

$$L_0 \, \frac{\delta F\{x,y\}}{\delta x(t)} + \frac{\delta^2 F\{x,y\}}{\delta x(t)\delta y(t)} = g(t)\, F\{x,y\} \tag{6.6}$$

Little is known about the solution of this equation[*].

A second possibility consists in expressing the moments of higher order in terms of the preceding lower-order moments by introducing certain, more or less arbitrary, assumptions. In this manner one obtains a closed, finite system of differential equations.

Assuming, for instance,[**] $z(t)$ and $u(t)$ to be approximately independent, one has

$$< z(t)\, u(t) > \, = \, < z(t) > < u(t) > \tag{6.7}$$

Eq. (6.3) then contains only the single unknown $<u>$ and reduces, if $< z(t) > \, = \, 0$, to

$$L_0 < u > \, = \, g(t) \tag{6.8}$$

A comparison with (4.6) shows, that the assumption is exact for white noise coefficients. <u>Keller</u> [8] has proved that it is generally admissible, provided the

---

[*] See, for instance, [10] , Chapter 3 .

[**] See also [9] .

stochastic part of the operator L is sufficiently
small. To show this, we consider the general equation
(6.1) with

$$L = L_0 + \varepsilon L_1 + \varepsilon^2 L_2 + O(\varepsilon^3) \qquad (6.9)$$

where ε is a small parameter, L_0 represents a deterministic operator, and L_1 , L_2 are stochastic
operators.

Let u_0 denote a particular solution of the
deterministic part of Eq. (6.1)

$$L_0 u_0 = g \qquad (6.10)$$

Putting, then,

$$u = u_0 + \varepsilon u_1 + \varepsilon^2 u_2 + O(\varepsilon^3) \qquad (6.11)$$

substituting into Eq. (6.1), using (6.9) and equating
to zero the coefficients of ε and ε^2 , one obtains

$$L_0 u_1 + L_1 u_0 = 0 , \quad L_0 u_2 + L_1 u_1 + L_2 u_0 = 0$$

Multiplication from the left with the inverse operator L_0^{-1} renders

$$u_1 = - L_0^{-1} L_1 u_0 , \quad u_2 = L_0^{-1} \left(L_1 L_0^{-1} L_1 - L_2 \right) u_0$$

Hence, the solution of (6.1) is, if terms in ε^2 are included,

$$u = u_0 - \varepsilon L_0^{-1} L_1 u_0 + \varepsilon^2 L_0^{-1}\left(L_1 L_0^{-1} L_1 - L_2\right) u_0 + O\left(\varepsilon^3\right) \qquad (6.12)$$

Averaging gives for the desired expectation

$$<u> = u_0 - \varepsilon L_0^{-1} <L_1> u_0 + \varepsilon^2 L_0^{-1}\left[<L_1 L_0^{-1} L_1> - <L_2>\right] u_0 + O\left(\varepsilon^3\right)$$

$$(6.13)$$

For a comparison with (6.3) we eliminate u_0

$$u_0 = <u> + \varepsilon L_0^{-1} <L_1> u_0 + O\left(\varepsilon^2\right)$$

$$= <u> + \varepsilon L_0^{-1} <L_1> <u> + O\left(\varepsilon^2\right)$$

and substitute into (6.13)

$$<u> = u_0 - \varepsilon L_0^{-1} <L_1> <u> + \varepsilon^2 L_0^{-1}\left[<L_1 L_0^{-1} L_1> - <L_1> L_0^{-1} <L_1> - <L_2>\right] <u> +$$

$$+ O\left(\varepsilon^3\right) \qquad (6.14)$$

We have now a differential equation for $<u>$, which contains u_0 only in additive form. A different form is obtained by multiplying (6.14) by L_0 :

$$\left\{L_0 + \varepsilon <L_1> + \varepsilon^2\left[<L_1> L_0^{-1} <L_1> - <L_1 L_0^{-1} L_1> + <L_2>\right]\right\} <u> = g + O\left(\varepsilon^3\right)$$

$$(6.15)$$

Frequently, $<L_1> = 0$. Eq. (6.15) then simplifies to

$$\left\{L_0 + \varepsilon^2\left[<L_2> - <L_1 L_0^{-1} L_1>\right]\right\} <u> = g + O\left(\varepsilon^3\right) \qquad (6.16)$$

Within the order $O(\varepsilon)$ we have, therefore, exactly

$$L_0 < u > = g \tag{6.17}$$

which is identical with Eq. (6.8), obtained above with the aid of more or less arbitrary assumptions.

Eq. (6.8) may be improved by closing the hierarchy equations at higher moments rather than after the first step. This will be shown for the second step [11] . Consider the equation of linear oscillations with parametric excitation

$$\ddot{y} + \left[1 + z(t)\right] y = 0 \tag{6.18}$$

Averaging renders

$$\left(\frac{d^2}{dt^2} + 1\right) < y > + < zy> = 0 \tag{6.19}$$

Instead of making the crude assumption $< zy > = < z >< y >$, we multiply by $z(s)$ as above, and average

$$\left(\frac{d^2}{dt^2} + 1\right) < z(s) y(t) > + < z(s) z(t) y(t) > = 0 \tag{6.20}$$

In this manner we could continue. However, if we wish to stop with (6.20), we close the system by neglecting the dependence between $z(s) z(t)$ and $y(t)$, and put

$$< z(s) z(t) y(t)> = < z(s) z(t) >< y(t) >$$

We have now a closed system of two equations

$$\left(\frac{d^2}{dt^2}+1\right)<y(t)>+<z(t)y(t)> = 0$$

$$\left(\frac{d^2}{dt^2}+1\right)<z(s)y(t)>+<z(s)z(t)><y(t)>=0$$

$$(6.21)$$

From the second equation we find, using the impulse response of the system,

$$<z(s)y(t)> = -\int_0^t \sin(t-\tau)<z(s)z(\tau)><y(\tau)>d\tau$$

where $<z(s)z(\tau)> = R_z(s-\tau)$ for stationary $z(t)$. Letting now $s=t$, and substituting into the first of Eqs. (6.21), we obtain the following integro-differential equation for the mean $<y(t)>$:

$$\left(\frac{d^2}{dt^2}+1\right)<y(t)> = \int_0^t \sin(t-\tau) R_z(t-\tau) <y(\tau)> d\tau \qquad (6.22)$$

At the present time, the solution of this equation is only known for certain limiting cases. For instance, if one assumes $z(t)$ to represent wide-band noise, then its correlation time is very short, and $R_z(\tau)$ decreases rapidly with increasing τ . Therefore,

$$\left| R_0 \right| \gg \left| R_1 \right| \gg \left| R_2 \right| \gg \cdots \qquad (6.23)$$

where

$$R_n = \int_0^\infty s^n R_{\dot{z}}(s)\, ds \qquad (6.24)$$

The right-hand side of Eq. (6.22) may now be expanded in a power series in terms of $s = t - \tau$,

$$\int_0^t \sin s\, R_{\dot{z}}(s) < y(t-s) > ds = \int_0^t \left[s - \frac{s^3}{6} + \ldots \right] R_{\dot{z}}(s) \left[<y(t)> - s\frac{d}{dt}<y(t)>+\ldots \right] ds =$$

$$= R_1 <y(t)> - R_2\, \frac{d<y(t)>}{dt} + \ldots$$

and one obtains for $<y(t)>$ the differential equation

$$\frac{d^2}{dt^2} <y(t)> + R_2 \frac{d}{dt} <y(t)> + (1-R_1)<y(t)> = 0 \qquad (6.25)$$

It can be seen immediately that stability of the mean is determined by the sign of R_2. For $R_2 \geq 0$ we have stability, for $R_2 > 0$ we have asymptotic stability while, for $R_2 < 0$, we have instability of the mean.

If $z(t)$ goes over into white noise, then $R_{\dot{z}}(s) = \delta(s)$, and $R_1 = R_2 = \ldots = 0$. We then obtain the form of Eq. (4.6).

7. Linear Differential Equations with Constant Random Coefficients.

The problem of stability simplifies considerably if the coefficients f_{ij} in the linear differential equations are random constants rather than random functions. In order to obtain conditions for almost sure asymptotic stability, one first computes the Hurwitz determinants, in particular the determinant of order $n-1$, known as Routh discriminant. These determinants are functions of the random coefficients. Therefore, the probability of asymptotic stability (all determinants positive) can be determined. For almost sure asymptotic stability this probability must be equal to one.

As an <u>example</u>, consider the following system of differential equations describing the motion of an airplane with automatic pilot, $[12]$:

$$\ddot{\varphi} + M\dot{\varphi} + k\varphi = -N\eta$$

$$\dot{\eta} = \alpha \psi \qquad\qquad (7.1)$$

$$\psi = \varphi + \beta\dot{\varphi} - \frac{\eta}{a}$$

All coefficients, with the exception of M , are deterministic positive constants. M is random with χ^2-probability distribution (cf. Table II of the

Appendix), with 4 degrees of freedom:

$$
p_M(x) = \begin{cases} \dfrac{1}{4} x e^{-x/2} & \text{for } x \geq 0 \\[2ex] 0 & \text{for } x \leq 0 \end{cases} \qquad (7.2)
$$

Since a, α β, k, N, are all positive, a necessary and sufficient condition for asymptotic stability of the system (7.1) is $R > 0$, where R is the Routh discriminant,

$$
\left(M + \frac{\alpha}{a}\right)\left(k + M\frac{\alpha}{a} + N\alpha\beta\right) - N\alpha - k\frac{\alpha}{a} > 0
$$

With the abbreviations

$$
\frac{1}{2}\left(\frac{\alpha}{a} + k\frac{a}{\alpha} + Na\beta\right) = A, \qquad A^2 - N(\alpha\beta - a) = B \qquad (7.3)
$$

the condition may be written as

$$
(M + A)^2 > B \qquad (7.4)
$$

The probability that this condition is satisfied, is given by

$$
P\left[(M + A)^2 > B\right] = \int_B^\infty q_{(M+A)^2}(y)\,dy \qquad (7.5)
$$

The roots of $y = (M + A)^2$ are

$$
M_1 = +\sqrt{y} - A, \qquad M_2 = -\sqrt{y} - A
$$

Hence, from Eq. (I - 2.42), taking (7.2) into consideration,

$$q_{(M+A)^2}(y) = \frac{p(\sqrt{y}-A)}{2|M_1+A|} + \frac{p(-\sqrt{y}-A)}{2|M_2+A|} = \begin{cases} \dfrac{\sqrt{y}-A}{8\sqrt{y}} \, e^{\frac{A-\sqrt{y}}{2}} & \text{if } y \geq A^2 \\ \\ 0 & \text{if } y < A^2 \end{cases} \qquad (7.6)$$

Eq. (7.5) then renders

$$P\left[(M+A)^2 > B\right] = \frac{1}{8} \int_C^\infty \frac{\sqrt{y}-A}{\sqrt{y}} \, e^{\frac{A-\sqrt{y}}{2}} \, dy = e^{\frac{A-\sqrt{C}}{2}} \left(1 + \frac{\sqrt{C}-A}{2}\right) \qquad (7.7)$$

where

$$C = B, \quad \text{if } B > A^2$$

$$C = A^2, \quad \text{if } B \leq A^2$$

We have $P < 1$ in the first case, while in the second case, $P = 1$. Therefore, the latter case corresponds to almost sure asymptotic stability.

8. The Problem of First Passage.

In the preceding discussions of stability we have required for a system to be stable that its motion, as caused by a short initial disturbance, remain bounded for <u>all time</u> $t > 0$. This condition is,

under circumstances, too strong for practical applica
tions. Frequently,it suffices to have the system
remain within a prescribed region of motion or ampli
tude for a finite time only or, in other words, to
have the rate of increase of the amplitudes sufficient
ly low. This consideration leads to the problem of
reaching a prescribed boundary, or of passing through
it for the first time.

        Let the behavior of the system be descri
bed, exactly or approximately, by a finite number of
generalized coordinates $q_1(t), q_2(t) \ldots q_n(t)$. In the
phase space spanned by the 2n coordinates $q_1(t) \ldots q_n(t)$,
$\dot{q}_1(t) \ldots \dot{q}_n(t)$, a position of equilibrium corresponds to
a fixed point, while, for a moving system, the image
point describes a curve.

        Now, let a region R in the phase space,
called "admissible region of motion", be bounded by
the hypersurface S , Fig. 5.2. As long as the image
point of the system moves within this region we call
its behavior stable. Once it leaves the region we
consider this to be equivalent to failure.

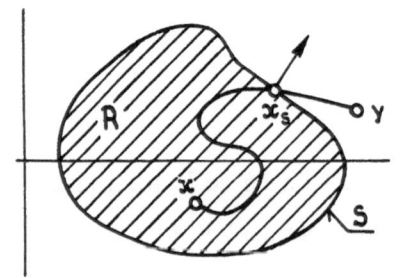

We denote by $U(t|x;0)$
the probability that the
image point, after having
started in $x = (q_1 \ldots q_n, \dot{q}_1 \ldots \dot{q}_n)$
at time $t = 0$, does not
leave the admissible region
R within time t U is

Fig. 5.2

thus a measure of the reliability of the system.

The probability

$$H(t|x;0) = 1 - U(t|x;0) \qquad (8.1)$$

represents the probability of first passage, i.e.,
the probability for the image point to pass through
boundary S for the first time within the time-interval $[0,t]$, after having started at x at time $t=0$.

Let $h(x_s;t|x;0)$ denote the probabi
lity density corresponding to (8.1), where x_s are the
phase coordinates of a point on the hypersurface S .
The quantity $h(x_s;t|x;0)\,dS\,dt$ then represents
the probability for the first appearance of the image
point on the surface element dS with outward velocity,
within the time-interval $(t,t+dt]$. Therefore

$$H(t|x;0) = \int_0^t d\tau \oint_S h(x_s;\tau|x;0)\,dS \qquad (8.2)$$

Now, if $p(x;t|x_0;0)$ denotes the transi-
tion probability of the process $x(t)$, assumed to be
Markov, one easily finds for $h(x_s;\tau|x;0)$ the
following integral equation, cf. Fig.5.2,

$$\left. \begin{array}{c} p(y;t|x;0) = \int_0^t d\tau \oint_S h(x_s;\tau|x;0)\,p(y;t|x_s;\tau)\,dS \\ x \in R, \quad x_s \in S, \quad y \notin R \end{array} \right\} (8.3)$$

which corresponds to the Chapman - Kolmogorov equation (I - 9.4). Also, an equation may be set up for
U ,[13], as the analogue of the Fokker - Planck equation (I - 9.16) of the process $x(t)$,

$$\frac{\partial U}{\partial t} = \sum_i f_i \frac{\partial U}{\partial x_i} + \sum_k \sum_r D_{kr} \sum_i \sum_j h_{ik} \, h_{jr} \frac{\partial^2 U}{\partial x_i \partial x_j} \qquad (8.4)$$

It differs from (I - 9.16) in that the right-hand
side is to be replaced by its adjoint expression. If
the "physical" Fokker-Planck equation is used the
corresponding changes have to be made in (8.4). For
initial condition one has

$$\lim_{t \to 0} U(t \,|\, x \,;\, 0) = 1 \qquad (8.5)$$

an or boundary condition

$$\lim_{x \to x_s} U(t \,|\, x \,;\, 0) = 0 \quad \text{for} \quad \sum_{i=1}^{2n} \dot{x}_i \, n_i > 0 \qquad (8.6)$$

since a crossing of the boundary surface S can only
occur if the image point has a velocity component
in the direction of the positive surface normal n ,
cf. Fig. 5.2.
          A solution to problems (8.4) - (8.6) is
not known at the present time. A numerical solution
of the integral equation (8.3) for a special case was
given by Parkus and Zeman [14] .

From (8.4) a differential equation may be obtained for the expectation $T \equiv <t>$ of first passage time. Putting

$$\frac{\partial H}{\partial t} = - \frac{\partial U}{\partial t} = \oint_S h(x_s;t|x;0)\,dS = \vartheta(t|x;0)$$

(8.7)

we note that the same differential equation (8.4) is valid for both U and ϑ, and we have

$$T = \int_0^\infty t\,\vartheta(t|x;0)\,dt$$

(8.8)

Replacing, therefore, U in Eq. (8.4) by ϑ, multiplying by t, and integrating, one obtains, taking the relation

$$\int_0^\infty t\,\frac{\partial \vartheta}{\partial t}\,dt = - \int_0^\infty \vartheta\,dt = -1$$

into consideration, the following differential equation for T

$$\sum_i f_i\frac{\partial T}{\partial x_i} + \sum_k \sum_r D_{kr} \sum_i \sum_j h_{ik}\,h_{jr}\frac{\partial^2 T}{\partial x_i\,\partial x_j} = -1$$

(8.9)

with $T = 0$ on the boundary S. Eq. (8.9) is known as the <u>Pontryagin</u> differential equation, [15]. Approximate solutions of this equation, obtained with the aid of the Galerkin method, are given by Bolotin, [15].

Approximation formulas for ϑ , for special cases, have been given in the literature. Let $n_a^+(t)$ denote the mean number of passages per unit time, with positive velocity, through $x = a$, cf. Eq. (IV-1.1),

$$n_a^+(t) = \int_0^\infty \eta \, p(a,\eta;t \mid x,\dot{x};0) \, d\eta \qquad (8.10)$$

Then, approximately, [16] ,

$$\vartheta(a;t \mid x,\dot{x};0) = n_a^+(t) \exp\left[-\int_0^t n_a^+(\tau) \, d\tau\right] \qquad (8.11)$$

If the process $x(t)$ has already become stationary, $n_a^+(t) = n_a^+ = \text{const}$, and Eq. (8.11) goes over into

$$\vartheta(a;t \mid x,\dot{x};0) = n_a^+ \, e^{-n_a^+ t} \qquad (8.12)$$

This is known as the exponential distribution, cf. Table II of the Appendix.

Literature .

[1] W.W.Bolotin: Kinetische Stabilität elastischer
    Systeme. VEB Deutscher Verlag der
    Wissenschaften, Berlin 1961.

[2] F. Kozin: A survey of stability of stochastic
    systems. Automatica 5 (1969), 95.

[3] F. Kozin: On almost sure stability of linear sys
    tems with random coefficients.
    J.Math.Phys. 42 (1963), 59.

[4] E.F. Infante:On the stability of some linear non-
    autonomous random systems. J.Appl.
    Mech. 35 (1968), 7.

[5] A.H.Gray, Jr.: Behavior of linear systems with
    random parametric excitation.
    J.Acoust.Soc.America 37 (1965), 235.

[6] P.W.U. Graefe: On the stabilization of unstable
    linear systems by white noise coef-
    ficients. Ing.-Arch.35 (1966), 276.

[7] J.L.Bogdanoff and F.Kozin: Moments of the output
    of linear random systems. J.Acoust.
    Soc.America 34 (1962),1063.

[8] J.B.Keller: Stochastic equations and wave propaga
    tion in random media. Proc.Symposia
    Appl. Math., vol.XVI, 1964,p.145.

[9] W.E. Boyce: A "dishonest" approach to certain

stochastic eigenvalue problems.
SIAM J. Appl. Math. 15 (1967), 143.

[10] M.J.Beran: Statistical Continuum Theories. Inter
science Publishers, New York 1968.

[11] J.M.Richardson: The application of truncated hi
erarchy techniques in the solution
of a stochastic linear differential
equation.Proc.Symposia Appl.Math.,
vol.XVI, 1964,p.290.

[12] Helga Bunke: Stabilität bei stochastischen Diffe
rentialgleichungssystemen. Z.ang.Math.
Mech. 43 (1963), 63.

[13] M.Kac: Probability theory. Proc. 1-st Symposium
on Engineering Applications of Random
Function Theory and Probability
(J.L. Bogdanoff and F. Kozin, editors).
J. Wiley and Sons, New York 1963, p.37.

[14] H.Parkus and J.L.Zeman: Some stochastic problems
of thermoviscoelasticity. Proc. IUTAM
Symposium on Thermoinelasticity.
Glasgow 1968 (under press)

[15] V.V. Bolotin:Statistical aspects in the theory of
structural stability. Proc.Int.Conf.
on Dynamic Stability of Structures.
(G.Herrmann, editor) Pergamon Press,
New York 1967, p. 67.

[16] J.R.Rice and F.P.Beer: First-occurrence time of
                high-level crossings in a continuous
                random process. J.Acoust.Soc.Ameri
                ca 39 (1966), 323.

C H A P T E R   VI

SYSTEM OPTIMIZATION .

In this chapter we consider a linear, open loop, deterministic transfer system, Fig. 6.1.

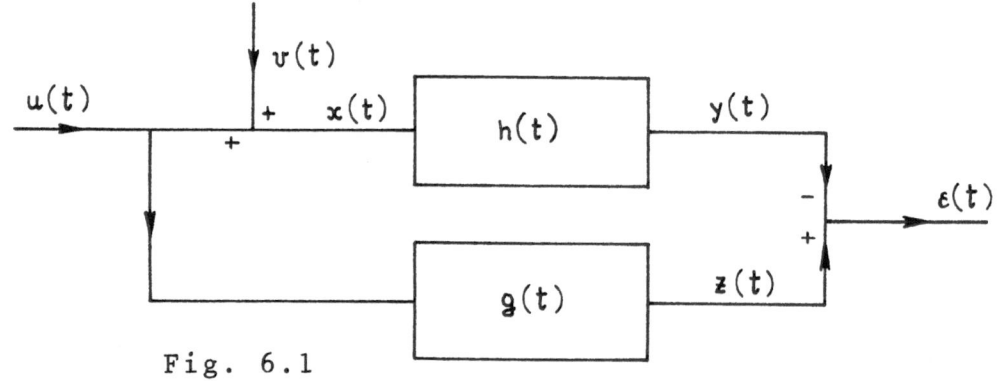

Fig. 6.1

The ideal system $g(t)$ transforms the undisturbed random test signal $u(t)$ into $z(t)$. The actual system $h(t)$, to be optimized, transforms the distorted signal $x(t)=$ $= u(t) + v(t)$, where $v(t)$ is random noise, into $y(t)$. We wish to have $y(t)$ "as close as possible" to $z(t)$ by minimizing the mean square error (dispersion D):

$$D = <\varepsilon^2(t)> = <\left[z(t) - y(t)\right]^2> = min.$$

Since we want to eliminate the perturbation noise $v(t)$ from $y(t)$ we speak of <u>filtering</u> or <u>smoothing</u>.

If $z(t) = u(t+\tau)$, we have the case of <u>extrapolation</u> or <u>prediction</u> where the future values of $u(t)$ are to be predicted from those of the past.

If

$$z(t) = \frac{d^k u(t+\tau)}{dt^k}$$

we have the case of <u>differentiation</u>.

1. The Wiener-Hopf Equation .

We assume the random processes $u(t)$ and $v(t)$ to be stationary with zero mean and known autocorrela tion and crosscorrelation. Furthermore, we assume that $x(t)$ has been observed through an unlimited period $(-\infty, t\,]$ preceding, and up to, time t .

Using the impulse response $h(t)$ of the sys tem we have for the dispersion, see (I - 5.20),

$$D = \left< \left[\int_0^\infty h(\tau)\, x(t-\tau)\, d\tau - z(t) \right]^2 \right> \qquad (1.1)$$

or, upon squaring,

$$D = \int_0^\infty h(s) \left[\int_0^\infty h(\tau)\, R_{xx}(s-\tau)\, d\tau - 2 R_{zx}(s) \right] ds + R_{zz}(0) \qquad (1.2)$$

$h(t)$ has to be determined such as to make D a minimum. To solve this variational problem we replace $h(t)$ by the expression

$$h(t) + \alpha \eta(t)$$

and require

$$\frac{\partial D}{\partial \alpha}\bigg|_{\alpha=0} = 0$$

whence

$$\int_0^\infty \eta(s)\left[\int_0^\infty h(\tau)\, R_{xx}(s-\tau)\, d\tau - R_{zx}(s)\right] ds = 0$$

From the fundamental lemma of the calculus of varia-
tions we get

$$\int_0^\infty h(\tau)\, R_{xx}(t-\tau)\, d\tau - R_{zx}(t) = 0 \qquad (1.3)$$

for $t \geqslant 0$, corresponding to the interval of integration of s .

Eq. (1.3) is known as the <u>Wiener-Hopf integral equation</u> for the optimal impulse response $h(t)$.
Due to the fact that it is valid for nonnegative t
only, its solution is not straightforward.

If (1.3) is substituted into (1.2) one obtains for the minimum dispersion

$$D_{min} = R_{zz}(0) - \int_0^\infty h(s)\, R_{zx}(s)\, ds \qquad (1.4)$$

2. Solution of the Wiener-Hopf Equation .

We first transform to spectral densities using Eqs. (I - 5.17),and introduce the frequency response

$$F (\omega) = \int_0^\infty h (\tau) e^{-i\omega\tau} d\tau$$

Eq. (1.3) then goes over into

$$\int_{-\infty}^{+\infty} \left[F(\omega) \, S_{xx} (\omega) - S_{zx}(\omega) \right] e^{i\omega t} d\omega = 0 \quad \text{for } t \geq 0 \quad (2.1)$$

To solve this integral equation for the optimal frequency response $F (\omega)$ we employ the following results from complex function theory:

Let $\Phi^+(\zeta)$ be a function without poles in the upper half complex plane $\eta \geq 0$, where $\zeta = \xi + i\eta$, and let $\Phi^-(\zeta)$ be a function without poles in the lower half plane $\eta \leq 0$. Then

$$\left. \begin{array}{ll} \int_{-\infty}^{+\infty} \Phi^-(\xi) e^{i\xi t} d\xi = 0 & \text{if } t < 0 \\[3em] \int_{-\infty}^{+\infty} \Phi^+(\xi) e^{i\xi t} d\xi = 0 & \text{if } t \geq 0 \end{array} \right\} \quad (2.2)$$

The proof may easily be given using the calculus of
residues and closing the integration path by semi-
circles, with $R \longrightarrow \infty$, in the lower and upper half-
planes, respectively. A condition on $\Phi\,(\zeta)$ is
appropriate behavior for $|\zeta| \rightarrow \infty$. Since it will be
assumed in the following that $\Phi\,(\zeta)$ can be represented
as a Fourier integral,

$$\int_{-\infty}^{+\infty} | \Phi\,(\xi) |^2 \, d\xi < \infty \qquad (2.3)$$

will be valid, and the condition will be satisfied.

The second of Eqs. (2.2) is identical with
Eq. (2.1). Hence, the function

$$\Phi^+(\xi) = S_{zx}(\xi) - F(\xi)\,S_{xx}(\xi) \qquad (2.4)$$

must not have any poles in the upper half-plane.

We assume the optimal system to be stable.
Therefore, its transfer function Y(s),

$$Y(s) = \int_0^{\infty} h\,(t)\,e^{-st}\,dt$$

cannot have poles with positive real parts of s . But,
since $F(\omega) = Y(i\omega)$, it follows that $F(\omega)$ cannot have
poles with negative imaginary parts of ω , $F(\omega) = F^-(\omega)$.
Eq. (2.4) may now be written as

$$\Phi^+(\xi) + F^-(\xi)\,S_{xx}(\xi) = S_{zx}(\xi) \qquad (2.5)$$

Spectral density $S_{xx}(\zeta)$ will, in general, have poles both in the upper and lower half-plane. In many instances, however, it is possible to represent S_{xx} as the product of a function $S^+(\zeta)$ and a function $S^-(\zeta)$, without poles in the upper and lower half-plane, respectively:

$$S_{xx}(\zeta) = S^+(\zeta) \, S^-(\zeta) \qquad (2.6)$$

This can be done in a particularly simple manner, if S_{xx} is the quotient of two polynomials[*].

Eq. (2.5) may now be written as

$$\frac{S_{zx}(\xi)}{S^+(\xi)} = \frac{\Phi^+(\xi)}{S^+(\xi)} + F^-(\xi) \, S^-(\xi) \qquad (2.7)$$

We have thus resolved the function $S_{zx}(\xi) \big/ S^+(\xi)$ into two parts, the first representing the value, along the real axis, of a function analytic in the upper half plane, and the second representing the value, along the real axis, of a function analytic in the lower half plane. Now, such a resolution may, however, also be obtained with the aid of the Fourier transform. Indeed, for a function $\Phi(\xi)$ satisfying (2.3),

$$\Phi(\xi) = \frac{1}{2\pi} \int_{-\infty}^{+\infty} e^{-i\xi t} dt \int_{-\infty}^{+\infty} \Phi(\omega) e^{i\omega t} d\omega = \frac{1}{2\pi} \left[\int_{-\infty}^{0-} e^{-i\xi t} dt + \int_{0+}^{\infty} e^{-i\xi t} dt \right] \int_{-\infty}^{+\infty} \Phi(\omega) e^{i\omega t} d\omega$$

---

[*] For a more general case, see [2] , p.136.

whence at once

$$\Phi^-(\xi) = \frac{1}{2\pi} \int\limits_{0+}^{\infty} e^{-i\xi t} dt \int\limits_{-\infty}^{+\infty} \Phi(\omega) e^{i\omega t} d\omega \\[2em] \Phi^+(\xi) = \frac{1}{2\pi} \int\limits_{-\infty}^{0-} e^{-i\xi t} dt \int\limits_{-\infty}^{+\infty} \Phi(\omega) e^{i\omega t} d\omega \Biggr\} \qquad (2.8)$$

For a proof it suffices to substitute Eqs. (2.8) into (2.2), and to use the delta function representation

$$\frac{1}{2\pi} \int\limits_{-\infty}^{+\infty} e^{i(t-\tau)\xi} d\xi = \delta(t-\tau)$$

Upon replacing now the right-hand side of Eq. (2.7) by its Fourier transform, and equating the ⁻parts on both sides one obtains

$$F(\omega) = \frac{1}{2\pi S^-(\omega)} \int\limits_{0+}^{\infty} e^{-i\omega t} \int\limits_{-\infty}^{+\infty} \frac{S_{zx}(\alpha)}{S^+(\alpha)} e^{i\alpha t} d\alpha \, dt \qquad (2.9)$$

This equation determines the optimal frequency response.

Literature .

[1] A. Papoulis: Probability, Random Variables and
        Stochastic Processes. Mc Graw-Hill
        Book Comp. New York 1965.

[2] A.A. Sweschnikow: Untersuchungsmethoden der Theorie
        der Zufallsfunktionen. Teubner,
        Leipzig 1965.

Additional material from *Random Processes in Mechanical Sciences,*
ISBN 978-3-211-81086-6, is available at http://extras.springer.com